INTRODUCTION TO CRYSTALLOGRAPHY

Donald E. Sands

UNIVERSITY OF KENTUCKY

D0139496

DOVER PUBLICATIONS, INC.
NEW YORK

Published in Canada by General Publishing Company, Ltd., 30 Lesmill Road, Don Mills, Toronto, Ontario.
Published in the United Kingdom by Constable and Company, Ltd., 3 The Lanchesters, 162–164 Fulham Palace Road, London W6 9ER.

Bibliographical Note

This Dover edition, first published in 1993, is an unabridged republication of the work first published by W. A. Benjamin, Inc. (in the "Physical Chemistry Monograph Series" of the "Advanced Book Program"), 1969 (corrected printing 1975).

Library of Congress Cataloging-in-Publication Data

Sands, Donald, 1929–
    Introduction to crystallography / Donald E. Sands.
        p.   cm.
    Includes index.
    ISBN 0-486-67839-3
    1. Crystallography. I. Title.
QD905.2.S26 1993
548—dc20   93-35539
                                                                  CIP

Manufactured in the United States of America
Dover Publications, Inc., 31 East 2nd Street, Mineola, N.Y. 11501

Preface

THE proliferation and the importance of the results of crystal structure analysis confront the chemist with the need to learn the language of crystallography. This book is an outgrowth of the opinion that the training of the undergraduate chemistry major can include more of this language than the memorization of a list of lattice types. At the same time, it would be unreasonable and impractical to expect all chemists to become experts in this specialized field. The purpose, therefore, is to treat the subject in a manner that will quickly and painlessly enable the nonspecialist to read and comprehend the crystallographic literature. It is hoped that this introduction may serve as a useful starting point for those students who wish to pursue the subject further.

The principal message is contained in the first four chapters. That is, these chapters supply the vocabulary of crystallography, and descriptions of crystal structures should be rendered intelligible by acquaintance with this language. In order not to discourage the general reader, the use of mathematics has been kept to a minimum. The decision to omit vector and matrix methods was made reluctantly, in order to reduce the prerequisites, but the elegance thus sacrificed is a luxury that some exceptionally competent crystallographers get along without.

On the other hand, proficiency in vector algebra is a useful substitute for an aptitude for three-dimensional visualization.

Chapters 5 and 6 attempt to show where the results come from. The problem is described formally as that of determining the coefficients in a Fourier series. This approach avoids a lengthy treatment of the physical theory of scattering, and the mathematical background required should not be much more than elementary calculus. Only a sampling of the methods and techniques of structure determination can be provided in the limited space of these chapters, and the reader interested in more detail is referred to one of the many excellent advanced treatises in the field.

The final chapter describes some simple structures, and the principles learned from familiarity with these can readily be extended to more complex cases. An admonition should be applied here (and in the previous chapters): this material should be understood rather than memorized.

Exercises of varying degrees of difficulty are distributed throughout the book. Which ones should be attempted will depend upon the level of understanding desired. As an aid to self-study, the solutions are given at the end of the book.

Even such a brief book has called upon the cooperation of many people. I especially wish to express my appreciation to my students, Mr. James A. Cunningham and Mr. Theodore Phillips, for critically reading the manuscript in an early form, and to my colleagues and family for tolerating my neglect of other duties. Thanks are also due Professor Walter Kauzmann for his helpful suggestions and comments at an early stage in the writing.

DONALD E. SANDS

Lexington, Kentucky
December 1968

Contents

Chapter 1

CRYSTALS AND LATTICES

Crystallography is concerned with the structure and properties of the crystalline state. Crystals have been the subject of study and speculation for hundreds of years, and everyone has some familiarity with their properties. We will concentrate on those aspects of the science of crystallography that are of interest to chemists. Our knowledge of chemistry will help us to understand the structures and properties of crystals, and we will see how the study of crystals can provide new chemical information.

1-1 Definition of a crystal

Crystals frequently have characteristic polyhedral shapes, bounded by flat faces, and much of the beauty of crystals is due to this face development. Many of the earliest contributions to crystallography were based on observations of shapes, and the study of morphology is still important for recognizing and identifying specimens. However, faces can be ground away or destroyed, and they are not essential to a modern definition of a crystal. Furthermore, crystals are often too small to be seen without a high-powered microscope, and many substances consist of thousands of tiny crystals (*polycrystalline*). Metals are

1

crystalline, but the individual crystals are often very small, and faces are not apparent. The following definition provides a more precise criterion for distinguishing crystalline from noncrystalline matter.

A crystal consists of atoms arranged in a pattern that repeats periodically in three dimensions.[1]

The pattern referred to in this definition can consist of a single atom, a group of atoms, a molecule, or a group of molecules. The important feature of a crystal is the periodicity or regularity of the arrangement of these patterns. The atoms in benzene, for example, are arranged in patterns with six carbon atoms at the vertices of a regular hexagon and one hydrogen atom attached to each carbon atom, but in liquid benzene there is no regularity in the arrangement of these patterns.

The fact that benzene is a liquid rather than a gas at room temperature is evidence of the existence of attractive forces between the molecules. In the case of benzene these are relatively weak van der Waals' forces, and thermal agitation keeps the molecules from associating into ordered clusters. If benzene is cooled below its freezing point of 5.5°C, the kinetic energy of the molecules is no longer sufficient to overcome the intermolecular attractions. The molecules assume fixed orientations and positions with respect to each other, and solidification occurs. As

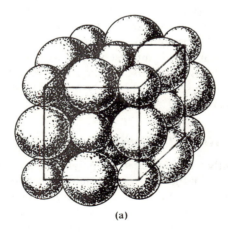

(a)

FIG. 1-1 (*a*)   *Portion of the sodium chloride structure showing the sizes of the ions, magnified about 10^8 times.*

[1] C. S. Barrett, *Structure of Metals*, 2nd ed., McGraw-Hill, New York, 1952, p. 1.

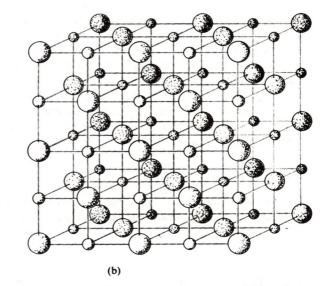

Cl⁻

Na⁺

(b)

FIG. 1-1 (*b*) *A model showing the geometrical arrangement of the sodium chloride structure.*

each molecule joins the growing solid particle, it is oriented so as to minimize the forces acting upon it. Each molecule entering the solid phase is influenced in almost exactly the same way as the preceding molecule, and the solid particle consists of a three-dimensional ordered array of molecules; that is, it is a crystal.[2]

Another example is afforded by a crystal of sodium chloride. The crystal contains many positive and negative ions held together by electrostatic attractions. The details of the arrangement depend upon the balancing of attractive and repulsive forces, which include both electrostatic and ionic size effects. The structure of sodium chloride is shown in Fig. 1-1, and further discussion will be given in Section 7-8. Each ion is surrounded by six ions of opposite charge, at the vertices of a regular octahedron, and the crystal structure represents an arrangement of these ions that leads to a potential energy minimum.

The point we want to emphasize here is that the formation of a solid

[2] This qualitative discussion is intended only to illustrate that periodicity is a natural consequence of the growth of a solid particle. It should not obscure the essential discontinuity of a phase change. See, for example, T. L. Hill, *Lectures on Matter and Equilibrium*, Benjamin, New York, 1966, p. 198.

particle naturally leads to crystallinity. There is a preferred orientation and position for each molecule to attach to the solid, and if the rate of deposition is slow enough to let the molecules attain this favored arrangement, the structure will fit our definition of a crystal. We have a pattern consisting of atoms or molecules. This pattern may be as simple as a single atom or it many consist of several molecules, each of which may contain many atoms. This entire pattern repeats over and over again, at regularly spaced intervals and with the same orientation, throughout the crystal.

1-2   Lattice points

Suppose we imagine a tiny creature wandering through the interior of a crystal. He stops at some point and closely examines his surroundings, and he carefully notes his position relative to the various atoms that

FIG. 1-2 *A two-dimensional periodic structure.*

FIG. 1-3 *Pick any point.*

constitute the pattern. He then walks in a straight line to an identical point in an adjacent pattern. For example, he might travel from one Na^+ ion to another Na^+ ion in sodium chloride or from the center of one ring to the center of another ring in benzene. When he arrives at this second point, there is absolutely nothing in his environment that will enable him to detect that he has moved at all. Furthermore, if he continues his walk without turning, he will come to another identical point when he has covered the same distance. Of course, the surroundings will look different near the surface, but in much of our discussion we will assume that the crystal contains so many repeating patterns that surface effects are quite negligible.

A useful two-dimensional analog of a crystal is an infinite wall covered with paper. The wallpaper pattern can be of any complexity, and the entire pattern repeats periodically in two dimensions. The array is actually periodic in the direction defined by any line connecting two identical points, but all of these directions can be described by taking vector sums of two arbitrary nonparallel base vectors.

Figure 1-2 shows a rather simple wallpaper pattern. To aid our discussions and calculations, it is convenient to choose some points and axes of reference. A system of reference points may be obtained by choosing one point at random (Fig. 1-3). All points identical with this point constitute the set of *lattice points* (Fig. 1-4). These points all have exactly the same surroundings, and they are identical in position relative to the repeating pattern or motif. This set of identical points[3] in two

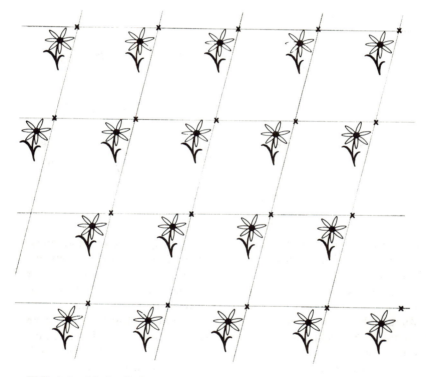

FIG. 1-4   *Mark all identical points. Connect points to form parallelograms.*

[3] We will frequently use the terms lattice point, identical point, and equivalent point interchangeably. Lattice points may be considered a special case of identical (or equivalent) points in that they are related to each other by lattice translations. All lattice points are equivalent to each other, but equivalent points are not necessarily lattice points.

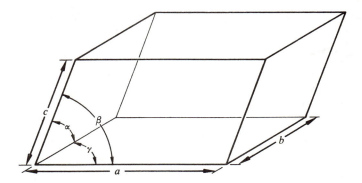

FIG. 1-5   *Six numbers specify the size and shape of a unit cell.*

dimensions constitutes a *net*. The term *lattice* or *space lattice* is frequently reserved for a three-dimensional distribution of points, and in one dimension the proper term is *row*.

1-3   Unit cells

If we now connect the lattice points by straight lines we can divide our two-dimensional space into parallelograms (Fig. 1-4). In three dimensions the space is divided into parallelepipeds. Repetition of these parallelepipeds by translation from one lattice point to another generates the lattice. The generating parallelepiped is called a *unit cell*. A unit cell is *always* a parallelepiped, and it is sort of a template for the whole crystal. If we know the exact arrangement of atoms within one unit cell, then we, in effect, know the atomic arrangement for the whole crystal. The process of determining the structure of a crystal consists, therefore, of locating the atoms within a unit cell.

The size and shape of a unit cell may be specified by means of the lengths a, b, and c of the three independent edges and the three angles α, β, and γ between these edges. These quantities are shown in Fig. 1-5. The angle α is the angle between b and c, β is between a and c, and γ is between a and b. These axes define a coordinate system appropriate to the crystal. In some respects it would be simpler to always use a Cartesian coordinate system, in which the three axes are equal in length and mutually perpendicular, but the advantages of a coordinate system based on the lattice vectors outweigh the simplicity of Cartesian geometry.

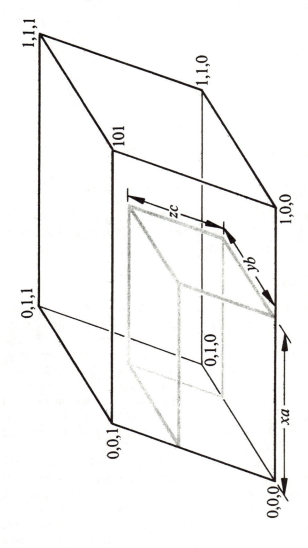

FIG. 1-6 *Location of a point with coordinates* x, y, z. *Numbers indicate coordinates of unit cell corners.*

EXERCISE 1-1   The density of NaCl crystals is 2.16 g/cc. Referring to Fig. 1-1, calculate the length of the edge of a unit cell of NaCl. (The unit cell in this case has $a = b = c$, $\alpha = \beta = \gamma = 90°$.)

1-4   Fractional coordinates

The location of a point within a unit cell may be specified by means of three fractional coordinates x, y, and z. The point x,y,z is located by starting at the origin (the point 0,0,0) and moving first a distance xa along the a axis, then a distance yb parallel to the b axis, and finally a distance zc parallel to the c axis (Fig. 1-6). If one of these coordinates is exactly 1, then the point is all the way across the unit cell, and if one of the coordinates exceeds 1, the point is in the next unit cell. For example, the point (1.30,0.25,0.15) is in the next unit cell to the right in Fig. 1-6. This point is equivalent to (0.30,0.25,0.15) since all unit cells are identical. It is thus apparent that a crystal structure can be entirely specified with fractional coordinates. One of the advantages of basing our coordinate system on lattice vectors is that two points are equivalent (or identical) if the fractional parts of their coordinates are equal. Also note that (−0.70,0.25,0.15) is equivalent to (0.30,0.25,0.15) since the x coordinates only differ by an integer; that is, equivalent points result when any integer is added to a coordinate.

1-5   Unit cell calculations

Calculations involving oblique coordinate systems are certainly more tedious than they would be if the axes were at right angles to each other, but compensation is provided by features such as the identity of the fractional coordinates of equivalent points in different unit cells. The following formulas will be useful.

The volume V of a unit cell is given by

$$V = abc(1 - \cos^2\alpha - \cos^2\beta - \cos^2\gamma + 2\cos\alpha\cos\beta\cos\gamma)^{1/2}$$

$$(1-1)$$

The distance l between the points x_1,y_1,z_1, and x_2,y_2,z_2 is

$$l = [(x_1 - x_2)^2 a^2 + (y_1 - y_2)^2 b^2 + (z_1 - z_2)^2 c^2$$
$$+ 2(x_1 - x_2)(y_1 - y_2)ab\cos\gamma + 2(y_1 - y_2)(z_1 - z_2)bc\cos\alpha$$
$$+ 2(z_1 - z_2)(x_1 - x_2)ca\cos\beta]^{1/2}$$

$$(1-2)$$

You should verify these formulas for the familiar case where $\alpha = \beta = \gamma = 90°$. Derivation of these formulas is accomplished easily by means of vector algebra.[4]

1-6   *Primitive and centered cells*

As is shown by Figs. 1-4, 1-7, and 1-8, the choice of a unit cell is not unique. Any parallelepiped whose edges connect lattice points is a valid unit cell according to our definition, and there are an infinite number of such possibilities. It is even permissible to have lattice points inside a unit cell (Fig. 1-8). In such cases there is more than one lattice

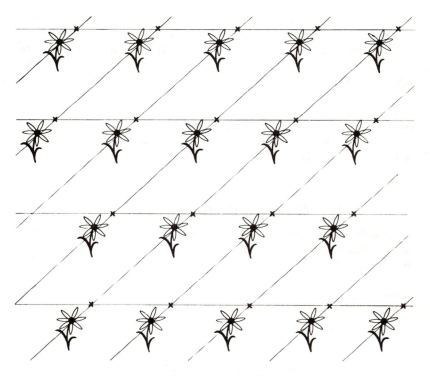

FIG. 1-7   *The choice of a unit cell is not unique.*

[4] Many crystallographic calculations are much easier when carried out by means of vector algebra. A brief introduction to vectors is given by D. A. Greenberg, *Mathematics for Introductory Science Courses*, Benjamin, New York, 1965.

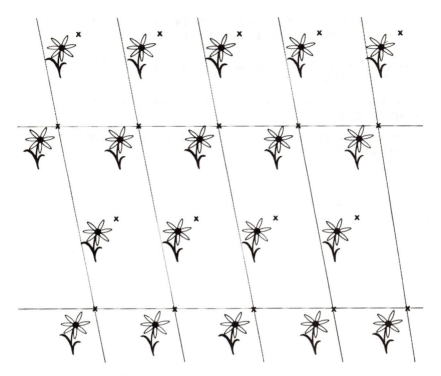

FIG. 1-8 *A unit cell need not be primitive.*

point per unit cell, and the cell is *centered*. A unit cell with lattice points only at the corners is called *primitive*. The centered cell in Fig. 1-8 is *doubly primitive*. A unit cell containing *n* lattice points has a volume of *n* times the volume of a primitive cell in the same lattice (see Exercise 1-2).

EXERCISE 1-2 A primitive unit cell has $a = 5.00$, $b = 6.00$, $c = 7.00$ Å; and $\alpha = \beta = \gamma = 90°$. A new unit cell is chosen with edges defined by the vectors from the origin to the points with coordinates $3,1,0$; $1,2,0$; and $0,0,1$.

(a) Calculate the volume of the original unit cell.

(b) Calculate the lengths of the three edges and the three angles of the new unit cell.

(c) Calculate the volume of the new unit cell.

(d) Calculate the ratio of this new volume to the volume of the original cell. How many lattice points does the new cell contain?

EXERCISE 1-3   A unit cell has dimensions $a = 6.00$, $b = 7.00$, $c = 8.00$ Å, $\alpha = 90°$, $\beta = 115.0°$, $\gamma = 90°$.

(a) Calculate the distance between the points $0.200, 0.150, 0.333$ and $0.300, 0.050, 0.123$.

(b) Calculate the distance between the points $0.200, 0.150, 0.333$ and $0.300, 0.050, -0.123$.

Chapter 2

SYMMETRY

2-1 Introduction

Some of the earliest studies of crystals were motivated by observations of their external symmetry. Snowflakes have hexagonal shapes; sodium chloride forms perfect cubes; and crystals of alum are frequently regular octahedra. Our study of crystallography will show us that these symmetrical shapes are manifestations of the internal structures of crystals. If the individual molecules have symmetry, then it is perhaps reasonable that they should pack together in a symmetrical array. There are some complicating factors, however, in the actual development of a crystal, and the arrangement of molecules is influenced by the availability of space as well as by the symmetry of the intermolecular forces. In the case of the very symmetrical benzene molecule, for example, it may turn out that six additional benzene molecules cannot approach closely from six equivalent directions without interfering with each other. Some of the molecules tilt to achieve a balance of attractive and repulsive forces, and the aggregate has less symmetry than the single molecule. In energy terms, the potential energy minimum is associated with a lower symmetry. It is also possible to have crystalline arrays where the symmetry is higher than that of the individual molecules, and we will encounter examples of this as we proceed. The important point here is that symmetry in crystals is a result of essentially the same factors that make crystallinity the natural state of a solid particle. If we have learned not to be surprised that solid matter is crystalline, then we must also accept the appearance of symmetry as an inevitable consequence of some rather simple laws. We will eventually see that crystallinity itself may be regarded as a special type of symmetry.

The utility of symmetry considerations extends beyond their application to crystals. For example, it is extremely useful to know that all six carbon atoms in a benzene molecule are identical (related by symmetry), and calculations pertaining to molecular vibrations or chemical bonding are vastly simplified by taking such symmetry into account. For this reason, we will devote this chapter to a general discussion of symmetry concepts and nomenclature, especially as applied to molecules. This material is indispensable to our later treatment of crystal geometry and symmetry; it is hoped that it will also provide some background for those who want to learn more about the role of symmetry in chemistry.[1]

2-2   Definition of symmetry

So far we have spoken of symmetry as though everyone knew what is meant by the term. This is probably true, at least qualitatively, but a definition will ensure a common understanding of the symmetry concept. *An object or figure is said to have symmetry if some movement of the figure or operation on the figure leaves it in a position indistinguishable from its original position.* That is, inspection of the object and its surroundings will not reveal whether or not the operation has been carried out.

2-3   Symmetry operations and elements of symmetry

A molecule of H_2O is shown in Fig. 2-1. The H—O—H angle is 104.5°, and the dotted line in the picture is the bisector of this angle. Suppose that the H_2O molecule is rotated 180° about the axis represented by the dotted line. The oxygen atom will be rotated 180°, but it will end up

[1] An excellent introduction to the noncrystallographic uses of symmetry is given by F. A. Cotton in *Chemical Applications of Group Theory*, Interscience, New York, 1963. A more advanced treatment may be found in R. M. Hochstrasser's *Molecular Aspects of Symmetry*, Benjamin, New York, 1966. See also J. E. White, *J. Chem. Educ.* **44**, 128 (1967).

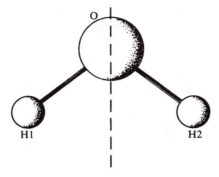

FIG. 2-1   *Water molecule.*

looking exactly as it did before the rotation. The two hydrogen atoms will be exchanged, so that now H(2) is on the left and H(1) is on the right. However, there is no possible way of telling the hydrogen atoms apart, since all hydrogens atoms are identical and in this molecule they have identical chemical environments. The 180° rotation has left the molecule in a position indistinguishable from its original position, so our definition of symmetry is satisfied. In H_2O the 180° rotation about the bisector of the H—O—H angle is a symmetry *operation*, and the rotation axis is a symmetry *element*. This particular symmetry element is designated by the symbol C_2 in the Schoenflies notation used extensively by spectroscopists, or simply by the symbol 2 in the Hermann–Mauguin or international notation preferred by crystallographers. Besides designating the symmetry element, the symbol C_2 (or 2) also implies the operation of rotation by 180°.

2-4   *Rotation axes*

A symmetry element for which the operation is a rotation of $360°/n$ is given the Schoenflies symbol C_n (or the Hermann–Mauguin symbol n). For example, the chloroform molecule ($CHCl_3$) has a C_3 axis. In this molecule (Fig. 2-2), the three chlorine atoms form an equilateral triangle, the carbon atom is directly above the center of this triangle, and the hydrogen atom is directly above the carbon atom. If the molecule is rotated about the axis defined by the C—H direction, it reaches identical orientations after every 120° of rotation, and there are a total of three equivalent orientations in a complete 360° turn.

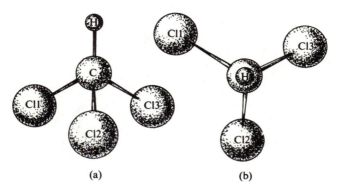

(a)   (b)

FIG. 2-2   *CHCl₃ molecule.*

2-5   *Mirror planes*

Rotation axes are not the only symmetry elements that molecules can possess, and both H_2O and $CHCl_3$ have mirror planes. If the dotted line in Fig. 2-1 represents a mirror perpendicular to the plane of the paper, one half of the molecule is just the mirror image of the other half. The $CHCl_3$ molecule has three mirror planes, corresponding to each of the three planes defined by the atoms H—C—Cl. If one of the chlorine atoms were replaced by a bromine atom, to form $CHCl_2Br$, the three-

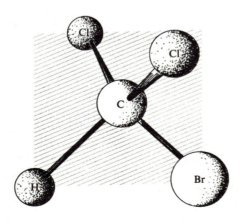

FIG. 2-3   *Mirror plane σ in CHBrCl₂.*

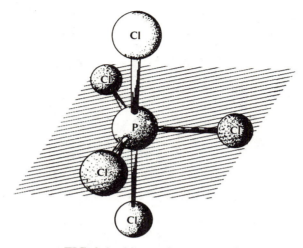

FIG. 2-4   *Mirror plane σ_h in PCl_5.*

fold symmetry would be destroyed, and the molecule would have only one mirror plane (Fig. 2-3).

The Hermann–Mauguin symbol for a mirror plane is *m*. In the Schoenflies notation the symbol σ is used with subscripts to indicate the orientation of the plane with respect to any rotation axes present. Thus, σ_h designates a *horizontal* mirror plane (Fig. 2-4) perpendicular

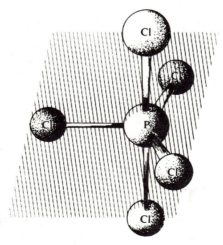

FIG. 2-5   *Mirror plane σ_v in PCl_5.*

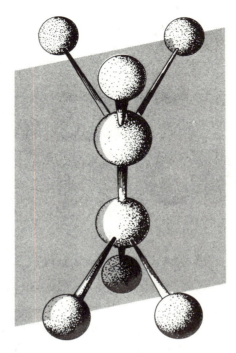

FIG. 2-6 *Mirror plane σ_d in staggered form of C_2H_6.*

to the principal rotation axis (the rotation axis of the highest order); σ_v is a *vertical* mirror plane (Fig. 2-5) which includes the rotation axis; and σ_d is a *diagonal* mirror plane (Fig. 2-6), which includes the principal rotation axis and bisects the angle between a pair of C_2 axes that are normal to the principal rotation axis. The C_2H_6 molecule has a C_3 axis; the form shown in Fig. 2-6 has three C_2 axes perpendicular to the C_3 axis, and there are three σ_d planes bisecting the angles between these C_2 axes.

2-6   Identity

Any direction in any object is a C_1 axis, since a 360° rotation merely restores the original orientation. This symmetry element is called the identity and is symbolized by C_1 or by 1. In Section 2-11 we will denote this element by E.

2-7   *Center of symmetry*

Another symmetry element that occurs frequently is a center of inversion or center of symmetry. This symmetry element is a point, and the operation consists of inversion through this point. A straight line drawn through the center of inversion from any point of the molecule will meet an equivalent point at an equal distance beyond the center. If the center of inversion is at the origin of the coordinate system, for any point with coordinates x, y, z, there is an identical point with coordinates $-x, -y, -z$. Figure 2-7 shows a molecule, *trans*-CHClBrCHClBr, which is centrosymmetric. The center of inversion is denoted by the Schoenflies symbol i or by the Hermann–Mauguin symbol $\bar{1}$.

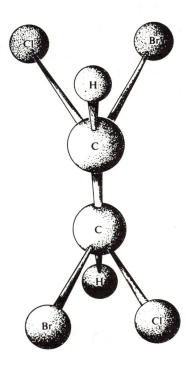

FIG. 2-7   *A molecule with a center of symmetry.*

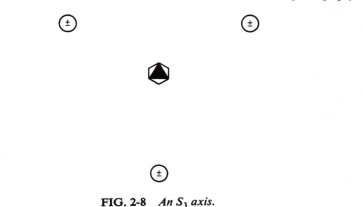

FIG. 2-8   *An S_3 axis.*

2-8   Improper rotation axes

The rotation axes we have discussed so far have been *proper rotation axes*. The only motion involved in the operation of a proper rotation axis is rotation by $360°/n$. Although the Schoenflies and Hermann–Mauguin systems use different symbols, the actual operations are identical. The operations in the two systems are quite different, how-

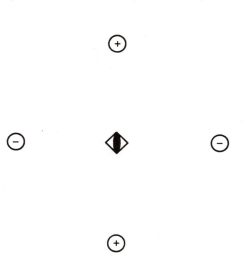

FIG. 2-9   *An S_4 axis* $(S_4 \equiv \bar{4})$.

ever, in the case of *improper rotations*. In the Schoenflies system, an improper rotation axis, S_n, is an axis of rotatory reflection, and the operation is a combination of rotation by $360°/n$ followed by reflection in a plane normal to the S_n axis. It is very important to understand that this definition does not necessarily imply the existence of C_n and σ_h individually. The operation is a *combination* of the motions of C_n and σ_h. An equivalent point is not generally reached after the C_n operation alone or after the σ_h operation alone, but only after both motions have been carried out. The effect of an S_3 axis is shown symbolically in Fig. 2-8. If we start with a point above the plane of the paper (labeled $+$), one application of S_3 generates a point rotated $120°$ and below the plane of the paper (labeled $-$). Successive applications of S_3 (S_3 applied to each newly generated point) give a total of six points, and Fig. 2-8 shows that, in this case, the presence of an S_3 axis implies both a C_3 axis and a

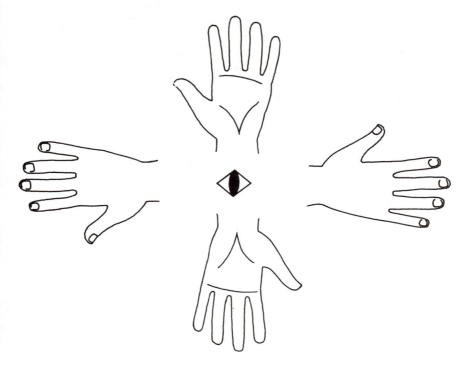

FIG. 2-10   *The effect of an S_4 axis.*

σ_h plane. An S_4 axis does not, however, imply either a C_4 axis or a σ_h plane. This is shown in Fig. 2-9, which obviously does not have a C_4 axis. Figure 2-10 also shows the effect of an S_4 axis, in which a right hand with the palm down is converted into a left hand with the palm up.

In the Hermann–Mauguin notation, an improper rotation axis is an axis of rotatory inversion. The operation here is a combination of rotation by $360°/n$ followed by inversion through a point, and the symbol is $\bar{n}$. Again, the operation consists of a combination of motions, and $\bar{n}$ does not necessarily imply the existence of either a proper rotation axis, n, or a center of inversion. A $\bar{3}$ axis is shown symbolically in Fig. 2-11. As in the case of S_3, a total of six points are generated by repeated application of the $\bar{3}$ operation. However, these points are distributed quite differently in the two cases. It is generally true that S_n and $\bar{n}$ are different unless n is a multiple of 4 (Fig. 2-12). It is easy to work out the effect of the S_n and $\bar{n}$ operations by means of diagrams such as Fig. 2-12, so no memory effort, other than the definitions of the operations, is required. It is of interest to observe that when n is odd $S_n = \overline{2n}$ and $\bar{n} = S_{2n}$. We thus have $S_1 = \bar{2} = \sigma(= m)$, $S_3 = \bar{6}$, $S_5 = \overline{10}$, and so on; and $\bar{1} = S_2 = i$, $\bar{3} = S_6$, $\bar{5} = S_{10}$, and so on. The symbol σ (or m) is always used in place of S_1 (or $\bar{2}$), and i is always used in place of S_2. We also note that when n is odd, S_n implies the presence of both C_n and σ_h, and $\bar{n}$ implies the presence of both n and $\bar{1}$.

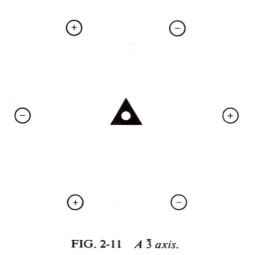

FIG. 2-11   *A $\bar{3}$ axis.*

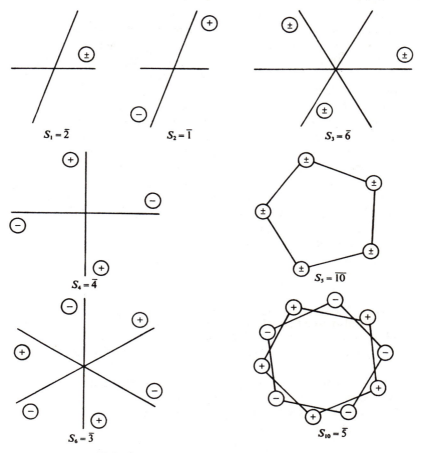

FIG. 2-12   *Comparison of S_n and $\bar{n}$ axes.*

2-9   *Point symmetry*

All of the symmetry operations we have discussed have the property that at least one point is not moved by the operation. All points on a proper rotation axis or on a mirror plane are stationary, and no new points are generated from these by the action of the operator. In the cases of the center of inversion and the improper rotations, there is a unique point that is left fixed. It is possible to conceive of symmetry operations that leave no point unchanged. If a set of points (or atoms) has such a symmetry element, and the operation is applied to produce a new set of points, then the new set must also contain the symmetry element (otherwise, the new set would be distinguishable from the first, in contradiction to our definition of symmetry). Operation by the new symmetry element will generate still another set of equivalent points, which will also contain the symmetry element. The result of the repeated application of newly generated symmetry elements is that an infinite number of equivalent points will be generated for each of our original points. It, therefore, follows that a finite molecule cannot have symmetry elements that do not leave at least one point fixed. For this reason, if we want to discuss the symmetry of molecules we need only the elements summarized in Table 2-1, which are referred to as elements of point symmetry. Crystals, however, can have symmetry elements that leave no point fixed (since we have agreed to regard our crystalline arrays as infinite), and translational symmetry will be dealt with at length in the following chapter.

2-10   Combinations of symmetry elements

We have already observed that molecules sometimes possess more than one symmetry element. Every molecule has a C_1 axis, and even the very simple water molecule has, in addition, a C_2 axis and two mirror planes (one of them is the plane of the molecule). We could describe the symmetry of a molecule by listing all of its symmetry elements. For CH_4 such a list would contain 24 entries, and SF_6 would require 48 entries. Even then, our symmetry description would not be clear unless we also explained how the symmetry elements are oriented with respect to each other.

We may have also observed that the symmetry elements of a molecule may not be mutually independent. For example, if two mirror planes intersect at right angles to each other, the line of intersection must be a twofold rotation axis. A simple proof of this is obtained by considering the changes in the coordinates of points as the symmetry operations are performed. An initial point at x, y, z is converted into $-x, y, z$ by a mirror plane perpendicular to the x axis. A mirror plane perpendicular to the y axis then converts $-x, y, z$ into $-x, -y, z$, and the point $-x, -y, z$ is just the result of operating on x, y, z by a twofold rotation axis parallel to the z direction (see Fig. 2-13). If we state that the H_2O molecule

TABLE 2-1 ELEMENTS OF POINT SYMMETRY

Type of element	Description of operation	Examples	
		Schoenflies symbol	Hermann–Mauguin symbol
Rotation axis	Counterclockwise rotation of $360°/n$ about axis	C_1, C_2, C_3, C_4	$1, 2, 3, 4$
Mirror plane	Reflection through a plane	σ	m
Identity	Rotation of $360°$ about any axis. All objects and geometric figures possess this element	$E = C_1$	1
Center of inversion (center of symmetry)	All points inverted through a center of symmetry	i	$\bar{1}$
Improper rotation axis (rotatory reflection axis)	Rotation of $360°/n$ followed by reflection in a plane perpendicular to the axis	S_1, S_2, S_3, S_4	
Improper rotation axis (rotatory inversion axis)	Rotation of $360°/n$ followed by inversion through a point on the axis		$\bar{1}, \bar{2}, \bar{3}, \bar{4}$

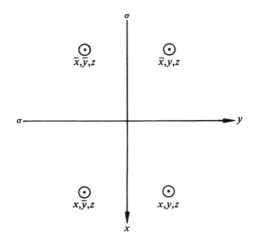

FIG. 2-13 *Two mutually perpendicular mirror planes generate a twofold rotation axis.*

possesses two perpendicular mirror planes, the symmetry of the molecule is completely described. We might alternatively denote the symmetry of this molecule by specifying that it has a C_2 rotation axis and a σ_v plane; the presence of C_2 and σ_v automatically imply the presence of another mirror plane.

2-11 Point groups

The collection of symmetry elements possessed by a molecule is called a *point group*. The word point indicates that at least one point of the molecule remains fixed under all of the symmetry operations. The word group means that some rather stringent conditions are satisfied:

A set of elements is called a group if there exists a law of combination, called multiplication, such that

1. the law of combination is associative,
2. there is an identity element in the set,
3. the inverse of every element is an element of the set, and
4. the product of any two elements is an element of the set.

The law of combination in our case consists of the successive application of two symmetry operations. For example, we might apply a C_2

operation followed by a σ_v operation. This combination of operations we write as $\sigma_v C_2$, where the order of the operations is read from right to left. Since the results of these two successive operations is another mirror plane, we can write $\sigma_v C_2 = \sigma_v'$. In this example $C_2 \sigma_v$ is also equal to σ_v'; that is, we get the same result if the σ_v operation is carried out first. This is not generally true, as shown by Fig. 2-14; point A is transformed into point D by the product $\sigma_v C_3$ (C_3 is a 120° counterclockwise rotation), but the product $C_3 \sigma_v$ transforms A into B. We have $\sigma_v C_3 = \sigma_v''$ and $C_3 \sigma_v = \sigma_v'$; since $\sigma_v C_3 \neq C_3 \sigma_v$, the group multiplication is said to be noncommutative.

Ordinary multiplication of numbers or algebraic quantities is commutative; in arithmetic $2 \times 3 = 3 \times 2$, and in elementary algebra $xy = yx$. In group theory, however, the word multiplication is broadly interpreted as implying two successive operations, and the order of the factors is important. If we write $\log \sin x$, this means compute the sine of the number x and then take the logarithm of the number $\sin x$. On the other hand, $\sin \log x$ means take the logarithm of x and then compute the sine of the number $\log x$. We find that $\log \sin x$ and $\sin \log x$ are generally not equal. If it suited our purposes we could develop an algebra of such operations and talk about the products $\log \times \sin$ and $\sin \times \log$. In group theory, this rather abstract type of multiplication is very useful, and diagrams, such as Fig. 2-14, aid in evaluating the products.

The law of combination is associative. If we consider the product of

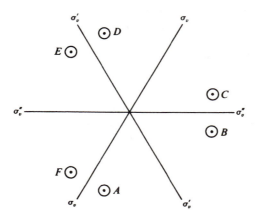

FIG. 2-14 *Point group* C_{3v}.

three elements, such as $\sigma_v'\sigma_v C_3$ in Fig. 2-14, we can write this product as either $\sigma_v'(\sigma_v C_3)$ or $(\sigma_v'\sigma_v)C_3$. Now, $\sigma_v C_3 = \sigma_v''$ and $\sigma_v'\sigma_v = C_3$, and our product is either $\sigma_v'\sigma_v''$ or $C_3 C_3$. The equality of these two products may be verified from Fig. 2-14 (they both convert point A into point E). We, therefore, see that while the order of the elements is important, they may be paired off in any manner.

The element C_1 is present in all of our symmetry groups, and C_1 serves as an identity element. Our symmetry groups are only special cases of collections of elements satisfying the group properties, and the symbol E is often used for the identity element, so in discussing group elements we shall use E instead of C_1. The identity is defined as an element E such that $EA = AE = A$, where A is any element of the group.

If A is any element of the group, there also exists an element A^{-1} such that $AA^{-1} = A^{-1}A = E$. The element A^{-1} is called the inverse of A. The inverse of C_2 is C_2, since $C_2 C_2 = C_2{}^2 = E$. The inverse of C_3 is $C_3{}^2 = C_3 C_3$, since $C_3 C_3{}^2 = C_3{}^2 C_3 = E$. The element $C_3{}^2$ may be regarded as either a 240° counterclockwise rotation or a 120° clockwise rotation.

The final property, which is that the product of any two elements of a group is also an element of the group, is called the law of closure. If A and B are elements of a group, then the product AB is an element of this group.

2-12 Group multiplication table

The properties of a group are given concisely by its multiplication table. The symmetry group of the H_2O molecule contains, as we have seen, the elements E, C_2, σ_v, and σ_v'. The number of elements is called the order of the group, so the order of this group is 4. The products of pairs of these elements are given in Table 2-2. The table is constructed by

TABLE 2-2   GROUP C_{2v} MULTIPLICATION TABLE

	E	C_2	σ_v	σ_v'
E	E	C_2	σ_v	σ_v'
C_2	C_2	E	σ_v'	σ_v
σ_v	σ_v	σ_v'	E	C_2
σ_v'	σ_v'	σ_v	C_2	E

TABLE 2-3   GROUP C_{3v} MULTIPLICATION TABLE

	E	C_3	$C_3{}^2$	σ_v	σ_v'	σ_v''
E	E	C_3	$C_3{}^2$	σ_v	σ_v'	σ_v''
C_3	C_3	$C_3{}^2$	E	σ_v'	σ_v''	σ_v
$C_3{}^2$	$C_3{}^2$	E	C_3	σ_v''	σ_v	σ_v'
σ_v	σ_v	σ_v''	σ_v'	E	$C_3{}^2$	C_3
σ_v'	σ_v'	σ_v	σ_v''	C_3	E	$C_3{}^2$
σ_v''	σ_v''	σ_v'	σ_v	$C_3{}^2$	C_3	E

listing each element across the top and at the left, and the products of pairs of elements are entered into the appropriate places. Although the elements of this particular group do commute, we have seen that this is not always the case, and we need to be careful about the order of the elements. We shall use the easily remembered convention that the element at the left in the table is also at the left in the product. Table 2-3 describes the group of order 6 illustrated in Fig. 2-14. This table tells us at a glance that $\sigma_v C_3 = \sigma_v''$ and $C_3 \sigma_v = \sigma_v'$, which are relationships that can be verified from Fig. 2-14.

The following properties of group multiplication tables are helpful in deriving the tables. Since $EA = AE = A$ for any element A, the first row and first column may be written down immediately. The positions of the E's in the table are readily obtained by considering the inverse of each element and using $AA^{-1} = A^{-1}A = E$. It is always true that each element of the group appears once and only once in each row and in each column.

EXERCISE 2-1   Verify that the numbers $1, -1, \sqrt{-1}, -\sqrt{-1}$ form a group where the law of combination is multiplication. Write the multiplication table.

EXERCISE 2-2   If $a^3 = 1$, write the multiplication table for the group with elements $a, a^2, 1$, where ordinary multiplication is the law of combination.

EXERCISE 2-3   Verify that the set of all integers $\ldots, -3, -2, -1, 0, 1, 2, 3, \ldots$ forms a group where addition is the law of combination.

EXERCISE 2-4   Derive the multiplication table for the group with elements $1, a, a^2, b, ab, a^2b$, using only the general properties of groups and the relationships $ba = a^2b$, $bab = a^2$, $aba = b$, $ba^2b = a$.

2-13   *Point group nomenclature*

A symmetry group may be designated very concisely by means of an abbreviated notation that gives sufficient information for deducing the detailed properties. In this chapter we will describe the older Schoenflies notation and supply some rules for determining point groups. The Hermann–Mauguin point group notation may be developed most logically after a discussion of crystal systems, so this topic will be deferred until Section 3-6.

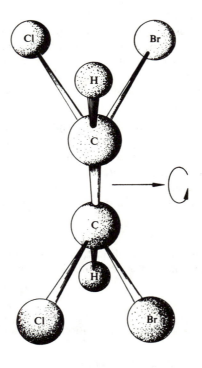

FIG. 2-15   *A molecule with C_2 symmetry.*

C_n GROUPS.   If the only elements of the group are a single C_n rotation axis and its powers C_n, C_n^2, ..., E, the group is called a cyclic group and is denoted by C_n.[2] The order of the group is n. A molecule with point symmetry C_2 is shown in Fig. 2-15; the only elements of the group are C_2 and E. The CH_3CCl_3 molecule has only C_3 symmetry if the chlorine atoms do not lie in the planes defined by the two carbon atoms and a hydrogen atom (Fig. 2-16a). The symmetry elements of the C_3 group are C_3, C_3^2, and E.

All elements of the C_n groups commute with each other; that is, $AB = BA$, where A and B are any two elements of the group. Groups with this property are called *Abelian*.

C_{nh} GROUPS.   A C_{nh} group has a mirror plane perpendicular to a C_n axis. When $n = 1$, the only symmetry elements are E and σ, and the

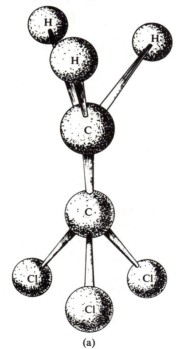

(a)

FIG. 2-16 CH_3CCl_3 *molecule.* (a) C_3 *symmetry* (continued on p. 32).

---

[2] No confusion should arise from using the symbol C_n to denote sometimes a group and sometimes a symmetry element or symmetry operation. The intended meaning will always be apparent from the context.

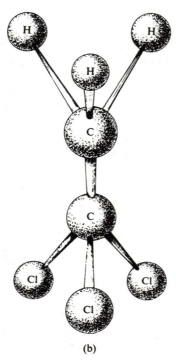

(b)

FIG. 2-16 CH_3CCl_3 molecule. (b) C_{3v} symmetry.

group is called C_s; the $CHBrCl_2$ molecule in Fig. 2-3 belongs to point group C_s. The trans configuration of $CHCl_2CHCl_2$, shown in Fig. 2-17, has C_{2h} symmetry. The symmetry operations of C_{2h} are C_2, σ_h, i, and E. The center of symmetry is not mentioned explicitly in this point group symbol, but its presence is a necessary result of the combination of a C_2 axis and a σ_h plane. Table 2-4 is the C_{2h} multiplication table.

TABLE 2-4 GROUP C_{2h} MULTIPLICATION TABLE

	E	C_2	σ_h	i
E	E	C_2	σ_h	i
C_2	C_2	E	i	σ_h
σ_h	σ_h	i	E	C_2
i	i	σ_h	C_2	E

C_{nv} GROUPS.   A C_{nv} group has a C_n axis and $n \sigma_v$ mirror planes. When $n = 1$, there is only a mirror plane, and $C_{1v} = C_{1h} = C_s$. The symmetry elements of C_{2v} are shown in Fig. 2-13, and molecules with this symmetry include H_2O (Fig. 2-1) and CH_2Cl_2 (Fig. 2-18). The symmetry elements of C_{3v} are shown in Fig. 2-14. Among molecules with C_{3v} symmetry are $CHCl_3$ (Fig. 2-2) and CH_3CCl_3 in both the staggered and eclipsed configuration (Figs. 2-16b and c). A C_{nv} group has order $2n$. Table 2-3 is the multiplication table for C_{3v}. The group $C_{\infty v}$ is of special interest since it is the symmetry group of all linear molecules that do not have a mirror plane perpendicular to the molecular axis. In such molecules (Fig. 2-19), the molecular axis is a C_∞ axis since any rotation leaves the molecule in an indistinguishable orientation and any plane that includes the axis is a mirror plane.

S_n GROUPS.   The elements of an S_n group are generated by application of an S_n axis. An S_1 axis is identical with a σ_h plane, and the correspond-

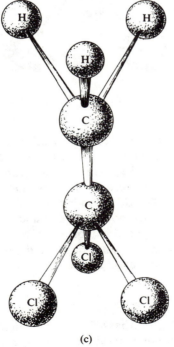

(c)

FIG. 2-16   *CH₃CCl₃ molecule. (c) C_{3v} symmetry.*

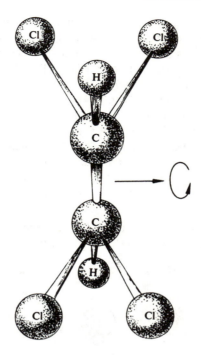

FIG. 2-17 $CHCl_2CHCl_2$, C_{2h} symmetry.

ing group is just C_s again. An S_2 axis is identical with a center of inversion, and the group, which possesses the elements E and i, is called C_i. When n is an odd number the groups S_n and C_{nh} are equivalent, and the C_{nh} notation is used. We, therefore, list separately only the S_n groups S_4, S_6, S_8, and so on. The S_6 group is sometimes called C_{3i} since its symmetry elements include both a C_3 axis and a center of inversion. Table 2-5 is the S_6 multiplication table.

The C_n, C_{nh}, C_{nv}, and S_n point groups are the only groups in which there is just one rotation axis. The cases we will now consider have two or more rotation axes.

D_n GROUPS.   A D_n group has n C_2 axes perpendicular to one C_n axis. The arrangements of points resulting from D_2 and D_3 symmetry are shown in Fig. 2-20. We can use ethane, C_2H_6, as an example of D_3

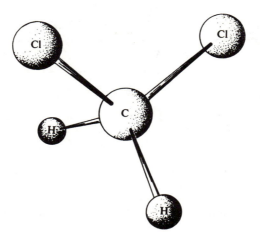

FIG. 2-18   *CH₂Cl₂, C₂ᵥ symmetry.*

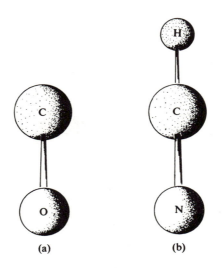

(a)   (b)

FIG. 2-19   *Linear molecules, group C∞ᵥ. (Compare Fig. 2-23.)*

TABLE 2-5   GROUP S_6 MULTIPLICATION TABLE

	E	S_6	C_3	i	C_3^2	S_6^5
E	E	S_6	C_3	i	C_3^2	S_6^5
S_6	S_6	C_3	i	C_3^2	S_6^5	E
C_3	C_3	i	C_3^2	S_6^5	E	S_6
i	i	C_3^2	S_6^5	E	S_6	C_3
C_3^2	C_3^2	S_6^5	E	S_6	C_3	i
S_6^5	S_6^5	E	S_6	C_3	i	C_3^2

symmetry provided that the molecule has neither the staggered nor the eclipsed configuration (Fig. 2-21a).

D_{nh} GROUPS.   In addition to the C_n axis and n C_2 axes of a D_n group, a D_{nh} group has a mirror plane perpendicular to the C_n axis. The eclipsed configuration of C_2H_6 (Fig. 2-21b) and 1,3,5-trichlorobenzene have D_{3h} symmetry. Table 2-6 is the D_{3h} multiplication table. The operations in Table 2-6 may be readily checked with the aid of Fig. 2-22. Note that the inverse of S_3 is S_3^5.

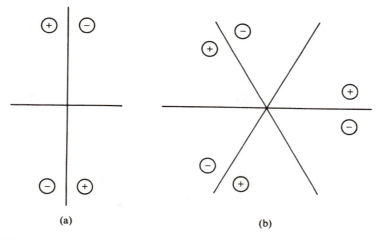

(a)   (b)

FIG. 2-20 D_n *symmetry. The C_n axis is normal to the paper.* (a) D_2, (b) D_3.

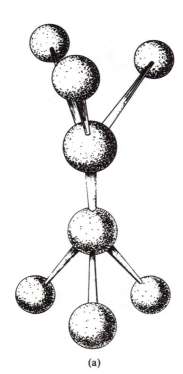

(a)

FIG. 2-21 *Three point groups for* C_2H_6. *(a)* D_3 *(continued on p. 38).*

(continued on p. 38)

TABLE 2-6 GROUP D_{3h} **MULTIPLICATION TABLE**

	E	C_3	$C_3^{\,2}$	C_2	C_2'	C_2''	σ_v	σ_v'	σ_v''	σ_h	S_3	$S_3^{\,5}$
E	E	C_3	$C_3^{\,2}$	C_2	C_2'	C_2''	σ_v	σ_v'	σ_v''	σ_h	S_3	$S_3^{\,5}$
C_3	C_3	$C_3^{\,2}$	E	C_2''	C_2	C_2'	σ_v''	σ_v	σ_v'	S_3	$S_3^{\,5}$	σ_h
$C_3^{\,2}$	$C_3^{\,2}$	E	C_3	C_2'	C_2''	C_2	σ_v'	σ_v''	σ_v	$S_3^{\,5}$	σ_h	S_3
C_2	C_2	C_2'	C_2''	E	C_3	$C_3^{\,2}$	σ_h	S_3	$S_3^{\,5}$	σ_v	σ_v'	σ_v''
C_2'	C_2'	C_2''	C_2	$C_3^{\,2}$	E	C_3	$S_3^{\,5}$	σ_h	S_3	σ_v'	σ_v''	σ_v
C_2''	C_2''	C_2	C_2'	C_3	$C_3^{\,2}$	E	S_3	$S_3^{\,5}$	σ_h	σ_v''	σ_v	σ_v'
σ_v	σ_v	σ_v'	σ_v''	σ_h	S_3	$S_3^{\,5}$	E	C_3	$C_3^{\,2}$	C_2	C_2'	C_2''
σ_v'	σ_v'	σ_v''	σ_v	$S_3^{\,5}$	σ_h	S_3	$C_3^{\,2}$	E	C_3	C_2'	C_2''	C_2
σ_v''	σ_v''	σ_v	σ_v'	S_3	$S_3^{\,5}$	σ_h	C_3	$C_3^{\,2}$	E	C_2''	C_2	C_2'
σ_h	σ_h	S_3	$S_3^{\,5}$	σ_v	σ_v'	σ_v''	C_2	C_2'	C_2''	E	C_3	$C_3^{\,2}$
S_3	S_3	$S_3^{\,5}$	σ_h	σ_v''	σ_v	σ_v'	C_2''	C_2	C_2'	C_3	$C_3^{\,2}$	E
$S_3^{\,5}$	$S_3^{\,5}$	σ_h	S_3	σ_v'	σ_v''	σ_v	C_2'	C_2''	C_2	$C_3^{\,2}$	E	C_3

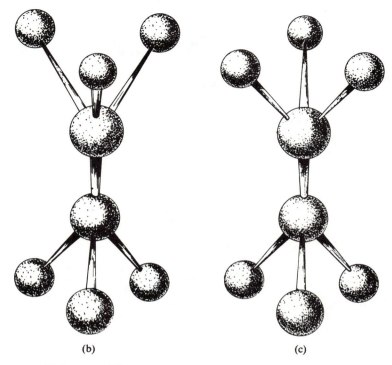

FIG. 2-21 *Three point groups for C_2H_6. (b) D_{3h}, (c) D_{3d}.*

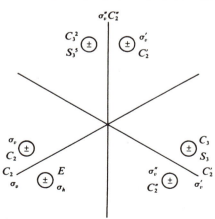

FIG. 2-22 *Symmetry operations of D_{3h}. Each point is labeled with the symmetry operation that generates it.*

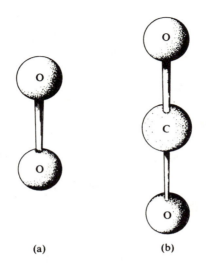

(a)   (b)

FIG. 2-23 *Linear molecules, group* $D_{\infty h}$. (*Compare Fig.* 2-19.)

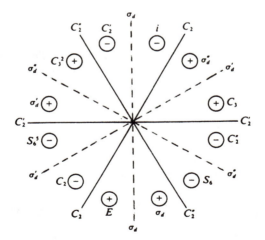

FIG. 2-24 *Symmetry operations of* D_{3d}.

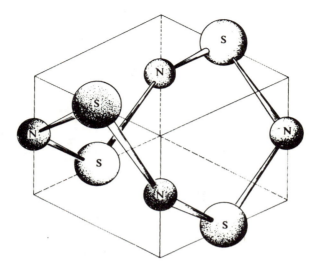

FIG. 2-25 N_4S_4, *a molecule with* D_{2d} *symmetry.*

The benzene molecule has D_{6h} symmetry. Linear molecules with a mirror plane perpendicular to the molecular axis have $D_{\infty h}$ symmetry (Fig. 2-23).

D_{nd} GROUPS.   A D_{nd} group is characterized by a C_n axis, by n C_2 axes perpendicular to the C_n axis, and by n vertical mirror planes that bisect the angles between the C_2 axes. The symmetry operations of D_{3d} are shown in Fig. 2-24, and the staggered configuration of C_2H_6 (Fig. 2-21c) has D_{3d} symmetry. The N_4S_4 molecule (Fig. 2-25) has D_{2d} symmetry.

CUBIC POINT GROUPS.   The next groups we will consider are the cubic point groups T, T_h, T_d, O, and O_h, which, in common with the cube, have four intersecting C_3 axes. The group T has all of the rotational symmetry elements of a regular tetrahedron. Figure 2-26 shows a regular tetrahedron inscribed in a cube. Each of the four body diagonals of the cube corresponds to a C_3 axis. In addition to these threefold axes, the group T has three C_2 axes parallel to the cube edges, bisecting opposite edges of the tetrahedron. The group T_h has a center of inversion, in addition to all of the elements of T. The group T_d has,

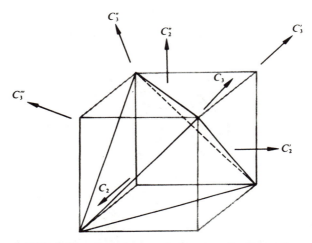

FIG. 2-26   *A regular tetrahedron inscribed in a cube.*

in addition to the four C_3 axes, three S_4 axes bisecting opposite edges of the tetrahedron. The symmetry group of tetrahedral molecules such as CH_4 (Fig. 2-27) is T_d; the group is of order 24.

The group O, of order 24, has all of the proper rotations of a regular octahedron (Fig. 2-28); these include four C_3 axes, three C_4 axes, and six C_2 axes. The group O_h has a center of inversion in addition to all

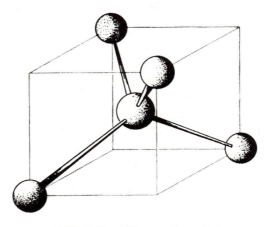

FIG. 2-27   *CH_4, symmetry T_d.*

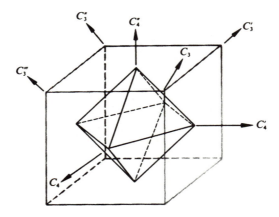

FIG. 2-28   *A regular octahedron inscribed in a cube. The C_3 and C_4 rotation axes are shown. Not shown are six C_2 axes parallel to the face diagonals of the cube.*

of the elements of O; O_h is of order 48 and is the symmetry group of octahedral molecules such as SF_6. (Fig. 2-29.)

ICOSAHEDRAL GROUPS.   Finally, we will mention the icosahedral groups I and I_h. The elements of group I are all of the rotations of a regular icosahedron (or a regular dodecahedron). These include six C_5 axes, ten C_3 axes, and fifteen C_2 axes. The group I_h has a center of inversion, in addition to all of the symmetry of group I. (See Fig. 2-30.)

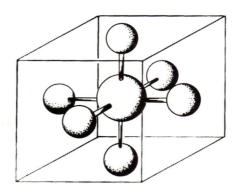

FIG 2-29 SF_6, *symmetry O_h.*

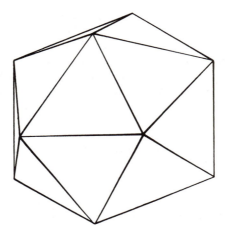

FIG. 2-30 *A regular icosahedron. There are six C_5 axes connecting opposite vertices, ten C_3 axes through opposite faces, and fifteen C_2 axes bisecting opposite edges.*

EXERCISE 2-5 Show that the order of group I is 60 by listing the total number of symmetry elements of each type ($C_5, C_5{}^2, C_5{}^3$, etc.).

2-14 Determination of point groups

The assignment of the point group of a molecule is based in great part on inspection. However, the following rules provide a systematic procedure.

1. Is the molecule linear? If so, the point group is $C_{\infty v}$ or $D_{\infty h}$.

2. Does the molecule have the high symmetry of the cubic point groups? The four threefold axes should be apparent, and a careful search for the other symmetry elements present should distinguish between T, T_h, T_d, O, and O_h. The abundance of symmetry of the icosahedral groups, I and I_h, should make these readily recognizable.

3. Having eliminated the highest symmetry groups, we now search for proper rotation axes. If there are none, the group is C_s, C_i, or C_1.

4. If there is more than one proper rotation axis, we try to select the one of highest order. A unique axis can be chosen except in the cases where there are three C_2 axes. Is this unique C_n axis actually an S_{2n} axis? If so, and if there is no other symmetry, the group is S_{2n}.

5. If the unique C_n axis is not an S_{2n} axis, or if there are other symmetry elements, we look for the presence of n C_2 axes perpendicular to C_n. If none are found, the group is C_n, C_{nv}, or C_{nh}, according to the presence of no mirror planes, a σ_v plane, and a σ_h plane.

6. If there are n C_2 axes perpendicular to C_n, the group is D_n, D_{nd}, or D_{nh}. It is D_{nh} if there is a σ_h plane; it is D_{nd} if there are n σ_d planes bisecting the angles between the C_2 axes; it is D_n if there are no mirror planes.

2-15 Limitation on combinations of symmetry elements

An interesting problem that arises is whether or not there exist symmetry groups other than those we have described. For example, is it possible to have a molecule that has two C_6 axes, or could we have a C_4 axis perpendicular to a C_3 axis? The answer is that there are no finite symmetry groups other than the ones we have discussed. Although there is nothing to stop us from carrying out the mathematical operations of combining perpendicular C_4 and C_3 axes, we would find that the products of elements continually lead to new elements, and the closure property could not be satisfied without an infinite number of elements.

Proofs of these statements are far beyond the scope of this book. However, it may be shown that the operation C_4C_3, where the C_3 axis is in the y direction and the C_4 axis is in the z direction, is identical with a rotation of $\cos^{-1}(-\frac{3}{4})$ about an axis from the origin through the point 1, -1, $1/\sqrt{3}$. The student who is adept in the algebra of vector transformations may want to verify this. The angle $\cos^{-1}(-\frac{3}{4})$ is not commensurate with 360°, and no matter how many times this rotation is carried out, the original position will not be reached again. Each application of the operator C_4C_3 thus generates a new point, and a molecule cannot have this symmetry combination without having all of the symmetry of a sphere. The only infinite groups allowed for nonspherical objects are $C_{\infty v}$ and $D_{\infty h}$, which apply to linear molecules.

EXERCISE 2-6 Prepare models that illustrate some of the cubic point groups by drawing the patterns shown in Fig. 2-31 on stiff paper or cardboard, cutting them out, folding on the dotted lines, and fastening with tape. Verify the presence of each of the required symmetry elements in the models. What are the point groups when there are no markings on the faces of (a) a cube, (b) a regular octahedron, (c) a regular tetrahedron?

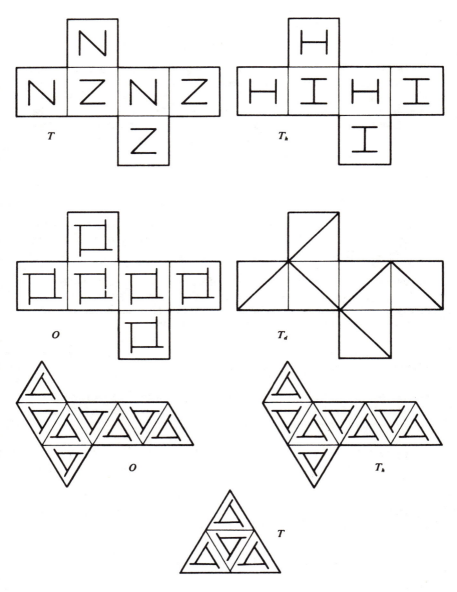

FIG. 2-31   *Patterns for construction of models illustrating some of the cubic point groups. See Exercise* 2-6.

EXERCISE 2-7   What is the point group of each of the three isomers of dichlorobenzene?

EXERCISE 2-8   What are the point groups of the linear molecules C_2H_2, C_2HCl, C_3O_2, and of the three isomers of $C_2H_2F_2$?

EXERCISE 2-9   The configuration of the molecule $Fe(C_5H_7O_2)_3$ is said to be octahedral because the oxygen atoms of the planar acetylacetonate groups are situated around the Fe^{3+} ion at the vertices of an octahedron. What is the point group of this molecule?

EXERCISE 2-10   The $AuCl_4^-$ ion is planar with the chlorine atoms at the vertices of a square. What is its point group?

EXERCISE 2-11   The following species have tetrahedral shapes: SO_2F_2, SO_4^{2-}, $Zn(NH_3)_4^{2+}$, $CFCl_3$, CF_2Cl_2. Give the point group of each. (Ignore hydrogen atoms.)

EXERCISE 2-12   Give the point group of each of the following molecules: (a) $MoCl_5$, Mo is at the center of a trigonal bipyramid; (b) Mo_2Cl_{10}, each Mo is surrounded by six Cl atoms at the vertices of an octahedron. One edge, defined by two Cl atoms, is shared by two octahedra.

EXERCISE 2-13   Sulfur forms S_8 molecules which have D_{4d} symmetry. Describe the structure of this molecule.

EXERCISE 2-14   The following species have octahedral shapes: $CrCl_6^{3-}$, $CrCl_5Br^{3-}$, $CrCl_4Br_2^{3-}$ (two isomers), $CrCl_3Br_3^{3-}$ (two isomers). Give the point group of each of these six ions.

EXERCISE 2-15   (a) The point group of the CO_3^{2-} ion is D_{3h}. Describe the structure of this ion. (b) The point group of NH_3 is C_{3v}. Describe the structure of the NH_3 molecule.

EXERCISE 2-16   Give the point group of each of the following: (a) N_2; (b) anthracene; (c) SF_5Cl; (d) cyclopropane.

Chapter 3

CRYSTAL SYSTEMS
AND GEOMETRY

Our definition of a crystal in Chapter 1 stressed its periodicity. We will subsequently consider the detailed structure of the repeating pattern in a crystal. However, our concern in this chapter will be the symmetry of the arrangement of atoms in the pattern. The determination of the symmetry groups compatible with a periodic structure will lead to a convenient classification of crystals.

3-1   Classification of unit cells

In Chapter 1 we pointed out that there are an infinite number of ways of choosing a unit cell for a given crystal structure. Of all the possible choices, there may be some that offer special advantages. For example, calculations of distances between atoms will be easier if the unit cell has 90° angles or if the axes are equal in length. Such considerations, there-fore, provide a possible criterion for selecting a favorable unit cell. A classification based on unit cell dimensions would not be entirely satisfactory, and one objection is that cell edges that appear to be equal might become unequal if the temperature were changed, unless there were some guarantee that the two directions experience the same thermal expansion. This effect might also cause the cell angles to deviate from 90°. Our classification scheme would depend upon our ability to detect these differences in dimensions, and it would be temperature dependent.

On the other hand, if we base our classification upon symmetry we will not run into this difficulty. If two directions in a crystal are equivalent by symmetry, they necessarily will have the same thermal expansion coefficients and they will remain equivalent with changing temperature unless there is a phase transition, as manifested by a discontinuity in other properties as well. Thermal expansion is only one of many physical properties of crystals that depend upon direction; other examples include electrical conductivity, magnetic susceptibility, elasticity, and optical properties.[1] Whereas the mathematical descriptions of these phenomena are frequently expressed in terms of Cartesian coordinates, their classification depends upon symmetry.

Our descriptions of crystal structures rely heavily on symmetry. For example, if a structure has a plane of symmetry, it is necessary only to list the positions of the atoms in half of the unit cell since the other half is generated by the reflection operation. There are relationships between the coordinates of symmetry related atoms, and if the choice of unit cell is dictated by symmetry these relationships are especially simple. This will be brought out more clearly in our development of sets of equivalent positions.

3-2 *Restrictions imposed by symmetry on unit cell dimensions*

We will, therefore, use symmetry in selecting a suitable unit cell. It will turn out that we will not after all have sacrificed the advantages attending a purely geometric choice; a unit cell chosen in accord with the symmetry elements will frequently have equal axes and 90° angles if such lattice vectors exist.

We will develop the lattice geometry by first considering a crystal with a twofold rotation axis. We will prove that the presence of this symmetry element guarantees that we can choose unit cell axes so one

---

[1] An intermediate mathematical treatment of the anisotropic properties of crystals is given by J. F. Nye, *Physical Properties of Crystals*, Oxford Univ. Press, London, 1957. A stimulating discussion of optical properties may be found in E. A. Wood, *Crystals and Light*, Van Nostrand, Princeton, New Jersey, 1963.

of them is perpendicular to the other two. Since the choice of the first lattice point is arbitrary, we might as well select a point on the twofold axis, and we will assign coordinates $0,0,0$ to this lattice point The direction of this twofold axis will be called the y direction, and any point on this axis will have coordinates $0,y,0$. Now, let us consider two other lattice points x',y',z' and x'',y'',z''. These x and z coordinates are referred to arbitrary axes; these axes are not necessarily lattice vectors, and for convenience we choose them normal to the y axis. As is apparent from Fig. 3-1, the twofold axis through the origin will generate lattice points with coordinates $-x',y',-z'$ and $-x'',y'',-z''$. Since there is no possible way of telling one lattice point from another, a twofold axis must pass through x',y',z', and careful consideration of Fig. 3-1 will show that the operation of this twofold axis on $-x'',y'',-z''$ generates a lattice point at $2x' + x'',y'', 2z' + z''$.

New lattice points can be generated by taking sums or differences of the coordinates of lattice points, so that $(x',y',z') + (x',y',z') = 2x',2y',2z'$ is a lattice point; so is $(x'',y'',z'') + (2x',2y',2z') = (2x' +$

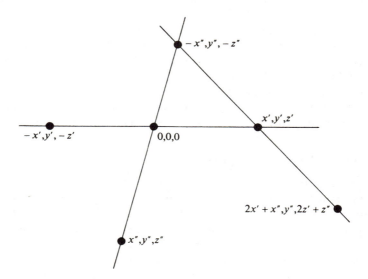

FIG. 3-1 *Effect of a twofold rotation axis perpendicular to the page. Given points x',y',z' and x'',y'',z'', points $-x',y',-z'$ and $-x'',y'',-z''$ are generated by the C_2 axis through $0,0,0$. The C_2 axis through x',y',z' generates $2x' + x'',y'',2z' + z''$ from $-x'',y'',-z''$.*

$x'', 2y' + y'', 2z' + z''$; and, finally, so is $(2x' + x'', 2y' + y'', 2z' + z'') - (2x' + x'', y'', 2z' + z'') = 0, 2y', 0$. We have thus proved that there exist lattice points along the y axis, and we must have $2y' = n$, where n is some integer. That is, we choose our b axis as the vector between adjacent lattice points in the y direction, and the coordinates of successive lattice points are $0,0,0$; $0,1,0$; $0,2,0$; and so on.

If n is an even number, y' is an integer, and the lattice point x', y', z' may be written x', m, z', where m is an integer. A new lattice point may be generated by the difference $(x', m, z') - (0, m, 0) = x', 0, z'$.

If n is an odd number, y' is a half-integer such as $\frac{1}{2}$, $\frac{3}{2}$, and $\frac{5}{2}$. In this case, the lattice point $2x', 2y', 2z'$ may be written as $2x', m, 2z'$, and $(2x', m, 2z') - (0, m, 0) = 2x', 0, 2z'$ is a lattice point.

Since either $x', 0, z'$ or $2x', 0, 2z'$ is a lattice point, and since any point with a zero y coordinate is in the plane of Fig. 3-1 perpendicular to the y direction, the vector from the origin to this point is a lattice vector perpendicular to the b axis. We can set $z' = 0$, and x' is either an integer or a half-integer. By similar reasoning, either $x'', 0, z''$ or $2x'', 0, 2z''$ is a lattice point. We have thus found two lattice vectors perpendicular to b, and we can call one of them a and one of them c. Therefore, as a consequence of the presence of a twofold rotation axis, it is possible to choose unit cell edges so that $\alpha = 90°$ and $\gamma = 90°$. This choice of axes is also possible if the symmetry operation is a mirror plane instead of a twofold axis; in this case the unique axis (usually labeled the b axis) is perpendicular to the mirror plane. Furthermore, point group C_{2h}, which contains both a C_2 axis and a perpendicular mirror plane, also requires the existence of two lattice vectors normal to the unique axis.

3-3 Crystal systems

These restrictions on lattice geometry characterize the monoclinic system. A crystal is said to be monoclinic if symmetry elements are present such that it is possible to pick a unit cell that has $\alpha = 90°$ and $\gamma = 90°$, with no other conditions on the dimensions and shape of the cell. The point groups that impose these, and only these, restrictions on the lattice vectors are C_2, C_s, and C_{2h}, and these are the monoclinic point groups (see Table 3-1). It is not sufficient to define the monoclinic system by merely stating $a \neq b \neq c$, $\alpha = 90°$, $\beta \neq 90°$, $\gamma = 90°$. It is the fact that $\alpha = \gamma = 90°$ *by virtue of symmetry* that characterizes the system as monoclinic.

There are two point groups that impose no restrictions on the lattice

TABLE 3-1 CRYSTALLOGRAPHIC POINT GROUPS

Crystal system	Schoenflies symbol	Hermann–Mauguin symbol	Order of group	Laue group
Triclinic	C_1	1	1	$\bar{1}$
	C_i	$\bar{1}$	2	
Monoclinic	C_2	2	2	$2/m$
	C_s	m	2	
	C_{2h}	$2/m$	4	
Orthorhombic	D_2	222	4	mmm
	C_{2v}	$mm2$	4	
	D_{2h}	mmm	8	
Tetragonal	C_4	4	4	$4/m$
	S_4	$\bar{4}$	4	
	C_{4h}	$4/m$	8	
	D_4	422	8	$4/mmm$
	C_{4v}	$4mm$	8	
	D_{2d}	$\bar{4}2m$	8	
	D_{4h}	$4/mmm$	16	
Trigonal	C_3	3	3	$\bar{3}$
	C_{3i}	$\bar{3}$	6	
	D_3	32	6	$\bar{3}m$
	C_{3v}	$3m$	6	
	D_{3d}	$\bar{3}m$	12	
Hexagonal	C_6	6	6	$6/m$
	C_{3h}	$\bar{6}$	6	
	C_{6h}	$6/m$	12	
	D_6	622	12	$6/mmm$
	C_{6v}	$6mm$	12	
	D_{3h}	$\bar{6}m2$	12	
	D_{6h}	$6/mmm$	24	
Cubic	T	23	12	$m3$
	T_h	$m3$	24	
	O	432	24	$m3m$
	T_d	$\bar{4}3m$	24	
	O_h	$m3m$	48	

vectors. These are C_1 and C_i and they characterize the triclinic system.

Point groups D_2, C_{2v}, and D_{2h} require that there exist three mutually perpendicular lattice vectors. It is, therefore, possible to select a unit cell with $\alpha = \beta = \gamma = 90°$. This is the orthorhombic system.

If the point group includes one (and only one) 4 or $\bar{4}$ axis (in Schoenflies notation C_4 or S_4), there exist vectors so that it is possible to choose $a = b$, $\alpha = \beta = \gamma = 90°$ with c parallel to the C_4 or S_4 axis. This is the tetragonal system.

The presence of a 6 or $\bar{6}$ axis (Schoenflies C_6 or S_3) characterizes the hexagonal system. If c is parallel to the sixfold axis, the hexagonal unit cell has $a = b$, $\alpha = \beta = 90°$, $\gamma = 120°$ (Fig. 3-2).

The presence of one (and only one) 3 or $\bar{3}$ axis (C_3 or S_6) denotes the trigonal system. Two types of lattice occur in the trigonal system, but a complete description of these lattices will be delayed until Section 3-7, so that we may make full use of the concept of centered lattices. For the present we state only that in one of these trigonal lattices a primitive unit cell may be chosen with $a = b$, $\alpha = \beta = 90°$, $\gamma = 120°$, whereas in the other trigonal lattice the primitive unit cell has $a = b = c$, $\alpha = \beta = \gamma$. When the primitive cell has $a = b$, $\alpha = \beta = 90°$, $\gamma = 120°$, the lattice is identical with the hexagonal lattice. When the primitive cell has $a = b = c$, $\alpha = \beta = \gamma$, the lattice is called rhombohedral, and the threefold symmetry axis is along the cell body diagonal.

If the point group includes four threefold axes, the system is cubic.

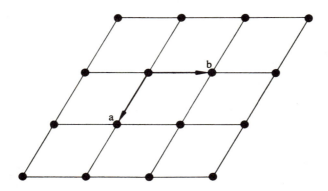

FIG. 3-2 *Distribution of lattice points in a hexagonal lattice. The 6 or $\bar{6}$ axis is perpendicular to the page. Nine unit cells are shown.*

FIG. 3-3 *The limitation of symmetry in a crystal. An n-fold rotation axis through A generates B′ from B. An n-fold rotation through D, at distance ma from A, generates C′ from C. B′C′ = la = ma − 2a cos 2π/n, where l, m, and n are integers. The only possible values of n satisfying this equation are 1, 2, 3, 4, and 6.*

It is possible to choose three equal axes at right angles to each other, and the four body diagonals of the unit cell cube will correspond to the threefold axes.

3-4   Limitations on symmetry in crystals

There are only seven crystal systems (some people count trigonal as part of the hexagonal system and so list only six systems), and there are only thirty-two crystallographic point groups. Although all point groups are permissible for isolated molecules, it is not possible for a crystal to have symmetries such as C_5 or D_{4d}. Figure 3-3 shows why only one-, two-, three-, four-, or sixfold rotation axes or axes of rotatory inversion are possible in a crystal. The n-fold rotation axes, perpendicular to the plane of the paper, generate $B′$ from B and $C′$ from C,

$$B′C′ = AD - 2a\cos\frac{2\pi}{n}$$

But $B′C′ = la$, where l is an integer, and $AD = ma$, where m is an integer.

$$la = ma - 2a\cos\frac{2\pi}{n}$$

$$l = m - 2\cos\frac{2\pi}{n}$$

$$\cos\frac{2\pi}{n} = \frac{m-l}{2}$$

The value of $\cos 2\pi/n$ must be between -1 and $+1$, so the allowed values are

$$\cos\frac{2\pi}{n} = -1, \qquad n = 2, \qquad \frac{2\pi}{n} = 180°$$

$$\cos\frac{2\pi}{n} = -\tfrac{1}{2}, \qquad n = 3, \qquad \frac{2\pi}{n} = 120°$$

$$\cos\frac{2\pi}{n} = 0, \qquad n = 4, \qquad \frac{2\pi}{n} = 90°$$

$$\cos\frac{2\pi}{n} = \tfrac{1}{2}, \qquad n = 6, \qquad \frac{2\pi}{n} = 60°$$

$$\cos\frac{2\pi}{n} = 1, \qquad n = 1, \qquad \frac{2\pi}{n} = 0° \text{ or } 360°$$

We, therefore, see that, although any symmetry is possible in a molecule, the point symmetry elements of a crystal are limited to one-, two-, three-, four-, and sixfold rotations and rotatory inversions (recall that $\bar{1}$ is a center of inversion and $\bar{2}$ is a mirror plane). A molecule may have $\bar{5}$ symmetry, and a crystal may be formed from such molecules. However, the symmetry of the environment of the molecule in the crystal cannot be $\bar{5}$ since $\bar{5}$ is not compatible with the requirements of translational symmetry. There are only thirty-two combinations of symmetry elements possible in a crystal, and these are the thirty-two crystallographic point groups listed in Table 3-1.

3-5   *Hermann–Mauguin notation*

In the Hermann–Mauguin system the point groups are designated by combinations of the symbols for symmetry elements. Some of the elements of the group are, therefore, immediately apparent from the symbol, and a few conventions make it possible to deduce the entire group structure. This system is preferred by crystallographers because it is easily extended to include translational symmetry elements and because it specifies the directions of the symmetry axes.

The Schoenflies and the Hermann–Mauguin symbols for the thirty-two crystallographic point groups are given in Table 3-1, and many of the features of the Hermann–Mauguin notation will be revealed by comparison with the Schoenflies symbols. The following summary should further clarify the meanings of the symbols:

1. Each component of a symbol refers to a different direction. The terms $2/m$, $4/m$, and $6/m$ are single components and refer to only one direction. In $4/mmm$, for example, the $4/m$ (read "four over m") indicates that there is a mirror plane perpendicular to a fourfold rotation axis.

2. The position of an m in a symbol indicates the direction of the normal to the mirror plane.

3. In the orthorhombic system, the three directions are mutually perpendicular. If we label our axes x,y,z, the symbol $mm2$ indicates that mirror planes are perpendicular to x and y, and a twofold rotation axis is parallel to z. The 2 in this case is redundant since we have seen that two perpendicular mirror planes inevitably generate a twofold axis. Note that such symbols as $m2m$ and $2mm$ correspond to renaming the axes.

4. If in the tetragonal system the 4 or $\bar{4}$ axis is in the z direction, the second component of the symbol refers to mutually perpendicular x and y axes, and the third component refers to directions in the xy plane that bisect the angles between the x and y axes.

5. In the trigonal and hexagonal systems, a second component in the symbol refers to equivalent directions (120° or 60° apart) in the plane normal to the 3, $\bar{3}$, 6, or $\bar{6}$ axis.

6. A third component in the hexagonal system refers to directions that bisect the angles between the directions specified by the second components.

7. A 3 in the second position always denotes the cubic system and refers to the four body diagonals of a cube. The first component of a cubic symbol refers to the cube axes, and a third component refers to the face diagonals of the cube.

The entries called Laue groups in Table 3-1 are the eleven centrosymmetric crystallographic point groups. If, for example, a center of symmetry is added to the list of elements of 422, or $4mm$, or $\bar{4}2m$, the result is $4/mmm$. The special significance of the Laue groups in crystallography will be explained in Section 5-21.

3-6 Bravais lattices

In selecting a unit cell based on symmetry elements, it may turn out that a nonprimitive, or centered, cell is obtained. In the triclinic system no symmetry restrictions occur, so a primitive cell can always be chosen. In the other crystal systems, however, centered cells are frequently encountered. In our development of the monoclinic system, based on Fig. 3-1, we started with a lattice point x', y', z' and proved the existence of a lattice point $0, 2y', 0$, where $2y' = n$. We also showed that either $x', 0, z'$ is a lattice point, or $2x', 0, 2z'$ is a lattice point, and the vector from the origin to this point is a lattice vector perpendicular to b. If both x' and y' are half-integers, the point $\frac{1}{2}, \frac{1}{2}, 0$ is a lattice point, and the unit cell is not primitive (Fig. 3-4). A primitive cell may, of course, be selected, but it would not be possible in this case to have b the unique axis and $\alpha = \gamma = 90°$. In order to preserve the advantages of a unit cell chosen on the basis of symmetry, a centered cell is chosen.

The unit cell described in this example is called C centered; the centering is on the C face, or the face of the unit cell bounded by the a and b axes. There are lattice points at $0, 0, 0$ and at $\frac{1}{2}, \frac{1}{2}, 0$. Points differing from these by l, m, n, where l, m, and n are integers, are also lattice points, of course, so that there are lattice points at $1, 0, 0;\ 0, 1, 0;\ 0, 0, 1;\ 1, 1, 0;$ $1, 0, 1;\ 0, 1, 1;\ 1, 1, 1;$ and $\frac{1}{2}, \frac{1}{2}, 1$; to list just a few of the infinite number of possibilities. (It would be well at this stage to recall the definition of a

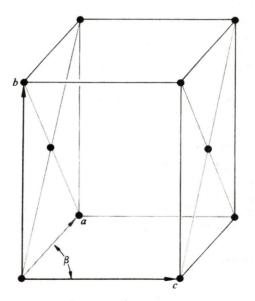

FIG. 3-4   *C-centered monoclinic unit cell. Lattice points are at 0,0,0 and at* $\frac{1}{2}, \frac{1}{2}, 0;\ \alpha = \gamma = 90°.$

lattice point: a lattice point can be any point in a crystal; the set of lattice points consists of all points identical except for translation.)

The C-centered monoclinic unit cell is shown in Fig. 3-4. It must be kept in mind that this diagram shows only a few lattice points and only one unit cell. The lattice point at $\frac{1}{2},\frac{1}{2},0$, which is in the center of the face bounded by a and b, is shared by two unit cells—by the cell shown in Fig. 3-4 and by the cell at its left. It is always true that a point in the center of a face is shared by two unit cells, and this must be taken into account in calculating the number of lattice points per unit cell in the whole crystal. In Fig. 3-4 we have lattice points at $\frac{1}{2},\frac{1}{2},0$ and $\frac{1}{2},\frac{1}{2},1$, but since each of these is shared by two unit cells the total contribution to one unit cell is $2 \times \frac{1}{2} = 1$ lattice point. Each of the eight vertices is shared by eight unit cells, so there is a contribution from the vertices of $8 \times \frac{1}{8} = 1$. Thus, our C-centered unit cell contains two lattice points. You should be very clear on this point. The fact that Fig. 3-4 contains ten lattice points and one unit cell does not mean that there are ten lattice points per unit cell. If we drew a second unit cell on the front of Fig. 3-4, so that the b and c axes of the two cells were in common, we would have sixteen lattice points and two cells in our picture. In order to assign correctly the number of lattice points, we must remember that we are depicting only a sample of an infinite lattice, divided into identical cells, in which corners are shared by eight cells, edges are shared by four cells, and faces are shared by two cells. It should now be obvious that a unit cell in which all faces are centered would contain $(6 \times \frac{1}{2}) + (8 \times \frac{1}{8}) = 4$ lattice points.

We have discovered that a lattice that is classified on the basis of symmetry as monoclinic may be primitive or it may be C-centered. Are other distinct types of centering possible? We could have A centering, but this differs from C centering only in our choice of names for the a and c axes, so this is not a distinct lattice type. Figure 3-5 shows two adjacent body-centered monoclinic unit cells. However, a new choice of axes results in a C-centered monoclinic unit cell. When we have a lattice with monoclinic symmetry, we will always be able to select either a primitive or a C-centered cell satisfying the monoclinic condition $\alpha = \gamma = 90°$. These are the only distinct lattice types consistent with monoclinic symmetry.

EXERCISE 3-1   Why is it not possible to have a unit cell that is centered on both the A and C faces?

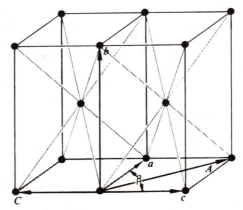

FIG. 3-5 *Two body-centered monoclinic unit cells. A C-centered cell may be selected with axes A and C = −c.*

EXERCISE 3-2 Show that *B*-centered monoclinic is not a distinct type of lattice.

The considerations we have applied to the monoclinic system may be extended to the other crystal systems. The result is that there are just fourteen of these space lattices. These were first deduced by Auguste

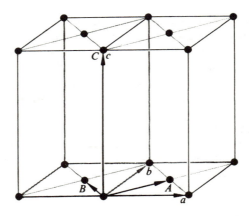

FIG. 3-6 *Two C-centered tetragonal unit cells. A primitive tetragonal cell is defined by the vectors A, B, C.*

Bravais in 1848, and they are frequently referred to as Bravais lattices. The fourteen Bravais lattices are listed in Table 3-2. The orthorhombic system includes four lattice types; besides the primitive and C-centered lattices, it is possible for a cell with orthorhombic symmetry to be either body centered (symbol I from the German *innenzentriert*) or face centered (symbol F, lattice points in the center of each face). The only distinct tetragonal lattices are primitive and body centered. The possibility of C-centered tetragonal is considered in Fig. 3-6. Similar analysis eliminates face-centering. However, it is not possible to describe a body-centered tetragonal crystal in terms of a primitive cell without sacrificing the geometrical advantages of the system, and if the arrangement of matter in a crystal is such that identical points form a body-centered tetragonal lattice, then the only logical description is in terms of these axes.

EXERCISE 3-3 Why is it not possible to have an orthorhombic cell that is I, A, B, and C centered all at once?

EXERCISE 3-4 Why is A-centered tetragonal not possible? How about A and B centering together?

3-7 *Distinction between trigonal and hexagonal systems*

The overlap between the trigonal and hexagonal systems was mentioned in Section 3-3. In this section we will further consider the distinction between these systems, and we will in particular establish the relationship between hexagonal and rhombohedral lattices.

As we brought out in Section 3-3, if the point group symmetry elements include a 6 or $\bar{6}$ axis, the system is hexagonal, and a primitive unit cell with $a = b$, $\alpha = \beta = 90°$, $\gamma = 120°$ may be chosen. If the axis of highest symmetry of the point group is a single 3 or $\bar{3}$ axis, the system is trigonal, and there are two possible lattices. If a primitive unit cell may be chosen such that $a = b$, $\alpha = \beta = 90°$, $\gamma = 120°$, the lattice is identical with the hexagonal case. The system is trigonal, as characterized by the presence of one 3 or $\bar{3}$ axis, but the distribution in space of lattice points is exactly the same as that resulting from hexagonal symmetry, and these are not two distinct lattice types. In both the trigonal and hexagonal systems, symmetry guarantees that cells with $a = b$, $\alpha = \beta = 90°$,

Tetragonal	4, 4̄, 4/m, 422, 4mm, 4̄2m, 4/mmm	$a = b$ $\alpha = \beta = \gamma = 90°$	
Trigonal *Rhombohedral*	3, 3̄, 32, 3m, 3̄m	$a = b = c$ $\alpha = \beta = \gamma$ or $a = b$ $\alpha = \beta = 90°$ $\gamma = 120°$	*R*
Hexagonal	6, 6̄, 6/m, 622, 6mm, 6̄m2, 6/mmm	$a = b$ $\alpha = \beta = 90°$ $\gamma = 120°$	*P*
Cubic	23, m3, 432, 4̄3m, m3m	$a = b = c$ $\alpha = \beta = \gamma = 90°$	*P* *I* *F*

TABLE 3-2 THE SEVEN CRYSTAL SYSTEMS AND FOURTEEN BRAVAIS LATTICES

Crystal system	Point group	Restrictions on axes or angles	Bravais lattices
Triclinic	$1, \bar{1}$	None	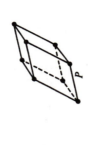 *P*
Monoclinic	$2, m, 2/m$	$\alpha = \gamma = 90°$	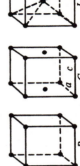
Orthorhombic	$222, mm2, mmm$	$\alpha = \beta = \gamma = 90°$	 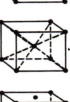

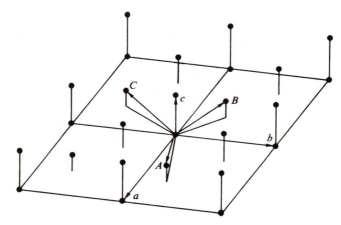

FIG. 3-7   *Rhombohedral lattice. The triply primitive hexagonal cell has axes a,b,c and lattice points at 0,0,0; $\frac{2}{3},\frac{1}{3},\frac{1}{3}$; $\frac{1}{3},\frac{2}{3},\frac{2}{3}$. Vectors from 0,0,0 to $\frac{2}{3},\frac{1}{3},\frac{1}{3}$; $-\frac{1}{3},\frac{1}{3},\frac{1}{3}$; and $-\frac{1}{3},-\frac{2}{3},\frac{1}{3}$ define the primitive rhombohedral cell.*

$\gamma = 120°$ can be chosen. It may be, however, that a cell for a trigonal crystal selected with these restrictions on the dimensions is not primitive. In these cases, it is possible to choose a primitive cell satisfying $a = b = c$, $\alpha = \beta = \gamma$. This lattice is called rhombohedral (symbol R). There are three equal axes inclined at equal angles with each other (Fig. 3-7). The trigonal threefold (or $\bar{3}$) axis in this case is along the body diagonal of the

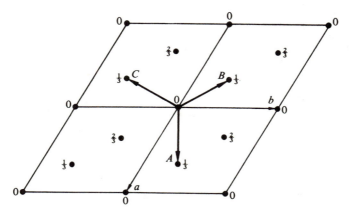

FIG. 3-8   *Rhombohedral lattice projected parallel to the threefold axis. Numbers next to points indicate z coordinates referred to hexagonal axes.*

cell (from point $0,0,0$ to point $1,1,1$). If the lattice is rhombohedral, it is still possible to choose a cell with hexagonal dimensions, but the cell will have 3 times the volume of the primitive cell. The relationship of the primitive rhombohedral cell to the triply primitive "hexagonal" cell is shown in Figs. 3-7 and 3-8. There are lattice points at $0,0,0$; $\frac{2}{3},\frac{1}{3},\frac{1}{3}$; and $\frac{1}{3},\frac{2}{3},\frac{2}{3}$ of the hexagonal cell. The three equal axes of the rhombohedral cell are given by vectors from the origin to the points $\frac{2}{3},\frac{1}{3},\frac{1}{3}$; $-\frac{1}{3},\frac{1}{3},\frac{1}{3}$; $-\frac{1}{3},-\frac{2}{3},\frac{1}{3}$. Figure 3-8 demonstrates a convenient method of depicting a complicated array of points; a projection of the arrangement is drawn, and the third coordinate of each point is written next to the point. Diagrams of this type will be used frequently in the following chapters.

EXERCISE 3-5   A rhombohedral unit cell has $a = 5.00$ Å, $\alpha = 75.0°$.

(a) Calculate the volume of the rhombohedral cell.

(b) Calculate the dimensions of the triply primitive hexagonal cell that may be chosen.

(c) Calculate the volume of the hexagonal cell from its dimensions and calculate the ratio of this volume to the volume obtained in (a).

EXERCISE 3-6   A rhombohedral unit cell has $a = 6.00$ Å, $\alpha = 60.0°$. Show that a face-centered cubic lattice may be chosen and calculate the dimension of the cubic cell. What is the criterion for deciding whether the crystal should be classified as cubic or as rhombohedral?

3-8   Crystal planes and indices

In Chapter 1 we mentioned that crystals frequently have polyhedral shapes bounded by flat faces and that observations of these faces played a crucial role in the historical development of crystallography. Although we preferred to define crystals on the basis of their internal structure, we must recognize that face development is a consequence of the periodicity of this internal arrangement of molecules or atoms. A study of the geometry of crystal planes will help us to understand the origin of crystal faces, but a reason that is more important for our purposes is that the description of these planes is essential to our interpretation of X-ray diffraction phenomena in Chapter 5.

A two-dimensional distribution of lattice points is shown in Fig. 3-9. We choose any two lattice points of this array, say A and B, and pass a plane through these points. (In three dimensions, three points not on a straight line are required to define a plane.) We now pass planes parallel to AB through every lattice point. We have generated a set of parallel equidistant planes, and these planes are all exactly alike. If we chose a lattice point at the center of a carbon atom in some organic crystal structure, then every lattice point would be at an identical carbon atom. The planes in Fig. 3-9 would then correspond to planes of identical carbon atoms; and these planes thus represent the stacking of layers of molecules.

There are an infinite number of such sets of parallel planes. The only restriction is that each plane must pass through at least two points in two dimensions or three noncollinear points in three dimensions. Each plane represents a layer of molecules, and each set of parallel planes represents a stacking of these layers. The faces of a crystal correspond to those planes that most favor the deposition of molecules. That is, the growing crystal adds molecules more easily on some planes than on others, and the corresponding crystal faces experience greater development.

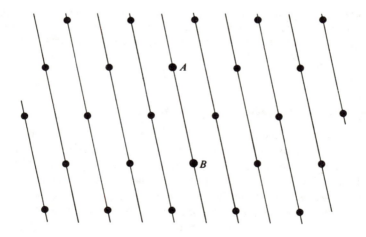

FIG. 3-9 *Two-dimensional distribution of lattice points. Pass a plane through points A and B. Pass planes parallel to AB through every lattice point. This generates a set of equivalent, parallel equidistant planes.*

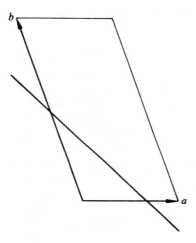

FIG. 3-10 *Planes with intercepts* $(\frac{2}{3}, \frac{1}{2}, \infty)$; *indices* (3 4 0).

The orientation of a set of parallel planes of a crystal may be specified by means of the intercepts of one of them on the three axes of the co-ordinate system, and it is the natural coordinate system of the crystal, with axes along the unit cell edges, that is used in our discussion. It is customary and convenient, however, to specify the orientations of crystal planes by means of their indices, which are proportional to the reciprocals of the intercepts. Figure 3-10 shows a plane that intercepts the a axis at $\frac{2}{3}$, the b axis at $\frac{1}{2}$, and the c axis at ∞ (this intercept on c merely means that the plane is parallel to c, so that c and the plane never intersect). The reciprocals of these intercepts are $\frac{3}{2}$, 2, and 0, and these three numbers can be used to characterize a plane. Now a plane with intercepts $\frac{1}{3}, \frac{1}{4}, \infty$ and indices 3,4,0 would be parallel with this plane, so if the orientation is all that interests us we can multiply the indices by a common factor so as to obtain integers for the indices. The plane shown in Fig. 3-10, therefore, has indices (3,4,0). As another example, a plane with intercepts $\frac{2}{7}, \frac{3}{5}, 1$ has indices proportional to $\frac{7}{2}, \frac{5}{3}, 1$, and if .we multiply each of these by 6 to get whole numbers, the indices are (21,10,6).

3-9   Law of rational indices

For some planes we might not be able to obtain whole numbers for the
indices. For example, if the intercepts are $1/\sqrt{2}, \frac{2}{3}, 1$, the indices will be
proportional to $\sqrt{2}, \frac{3}{2}, 1$, and there is no single factor that will convert
all three of these numbers into integers. However, as a result of the
periodicity of crystals, the only planes that are important in crystals
are those for which the indices are rational numbers (ratios of integers).
This is the law of rational indices, first deduced by Hauy in 1784 from
studies of crystal faces. Crystal faces can thus be described by means of
three indices that are whole numbers, and these, in fact, are always small
whole numbers for naturally growing crystals. The planes important to
our treatment of X-ray diffraction in Chapter 5 also have indices that
are integers, because planes with irrational intercepts do not constitute
sets of identical, parallel, equidistant planes.

The three integers describing the orientation of a plane are called
Miller indices, and the symbols h, k, and l are used for them.

EXERCISE 3-7   Write the Miller indices of the planes with intercepts (a)
$\frac{1}{2}, \frac{2}{3}, 1$; (b) $\infty, 1, \frac{2}{3}$; (c) $\frac{2}{3}, \infty, \frac{1}{6}$; (d) $\frac{1}{3}, \frac{2}{3}, \infty$; (e) $\frac{1}{6}, \frac{1}{3}, \infty$.

We now show that planes with rational intercepts pass through lattice
points. We will restrict ourselves to planes the indices of which are
relatively prime numbers; that is, the indices have no common factors.
We will, thus, concentrate at present on planes such as (100) and (123)
rather than (200) and (369). This is a logical restriction in classical
crystallography, since only the orientation of a crystal face has significance, and not how far it is from some arbitrarily defined origin. In
Section 5-7, we will find it convenient to remove this restriction.

Without loss of generality, the plane (hkl) can be described by the
equation

$$hx + ky + lz = 1 \qquad (3\text{-}1)$$

This equation has solutions where x, y, z are integers, provided that
h, k, l have no common factor. For example, the equation

$$21x + 10y + 6z = 1$$

is satisfied by such points as $1, -2, 0$ or $3, -5, -2$; that is, these points lie
on the plane (21 10 6). However, points for which the coordinates are

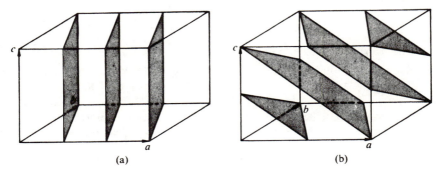

(a)   (b)

FIG. 3-11   (a) Sets of (310) planes; (b) sets of (212) planes.

integers are lattice points. Therefore, every plane (hkl) with relatively prime indices passes through a set of lattice points. Since all lattice points are identical, planes identical in all respects, including orientation, must pass through all lattice points. Figure 3-11 shows some members of the sets of (310) and (212) planes.

EXERCISE 3-8   Sketch, in cubic unit cells, planes with the following indices: (a) (100); (b) (120); (c) (111); (d) (11$\bar{1}$) (Note: $\bar{1}$ is a compact way of writing −1.)

3-10   Interplanar spacings

In Chapter 5 we will need to know how the interplanar spacing d for a set of parallel planes is related to the Miller indices and the unit cell dimensions. This spacing is the perpendicular distance between adjacent planes of the set. When the unit cell axes are mutually perpendicular, the interplanar spacing is easily derived by means of the formula from geometry for the distance from a point to a plane. The result

$$\frac{1}{d} = \left(\frac{h^2}{a^2} + \frac{k^2}{b^2} + \frac{l^2}{c^2}\right)^{1/2} \tag{3-2}$$

is applicable to orthorhombic, tetragonal, and cubic unit cells. When the axes are not mutually perpendicular, the derivation is best accomplished by means of vector algebra. The following formula applies to the most general case:

$$d = V[h^2b^2c^2 \sin^2 \alpha + k^2a^2c^2 \sin^2 \beta + l^2a^2b^2 \sin^2 \gamma$$
$$+ 2hlab^2c(\cos \alpha \cos \gamma - \cos \beta) + 2hkabc^2(\cos \alpha \cos \beta - \cos \gamma)$$
$$+ 2kla^2bc(\cos \beta \cos \gamma - \cos \alpha)]^{-1/2} \qquad (3\text{-}3)$$

where V is the unit cell volume given by Eq. (1-1). You should verify that Eq. (3-3) reduces to Eq. (3-2) when the angles are all 90°.

EXERCISE 3-9   Use Eqs. (3-3) and (1-1) to derive formulas for $1/d$ for (a) a monoclinic crystal; (b) a hexagonal crystal; (c) a rhombohedral crystal.

EXERCISE 3-10   A monoclinic crystal has $a = 5.00$ Å, $b = 6.00$ Å, $c = 8.00$ Å, $\beta = 115.0°$. Calculate d for the (101) planes by (a) drawing a diagram with the planes shown and determining d from the geometry of the figure; (b) applying Eq. (3-3).

EXERCISE 3-11   Repeat Exercise 3-10 for the (10$\bar{1}$) planes.

Chapter 4

SPACE GROUPS AND
EQUIVALENT POSITIONS

In the preceding chapters we developed the concepts and terminology of symmetry as applied to molecules, and we saw how symmetry considerations lead to a logical classification scheme for crystal lattices. A complete symmetry classification of crystals requires that we consider translations as well as operations of point symmetry.

4-1 Translational symmetry

Our definition in Chapter 2 referred to a symmetry operation as some movement after which no change could be detected in an object. The symmetry operations included in our previous discussions have had the common property that at least one point of the object was not moved by the operation. In the groups consisting of combinations of such elements there is also at least one point that remains fixed, and these groups are, therefore, called point groups. This was discussed in Chapter 2, where we reached the conclusion that the point groups are the only combinations of symmetry elements applicable to finite molecules. Other symmetry elements would lead to infinite molecules. A crystal, however, may be regarded as an infinite molecule; at least we have combinations of atoms that are repeated over and over throughout three-dimensional space. The lattice translations themselves satisfy our definition of symmetry operations, since the crystal is indistinguishable after such translations. The symmetry groups appropriate to crystals, therefore, contain infinite numbers of elements, and the lattice translations are included among these. There are two other types of symmetry element that result from combining the motions of rotations or reflections with the translatory symmetry of the lattice.

4-2　　Screw axes

The operation that characterizes a screw axis, for which the symbol is n_p, is a rotation of $2\pi/n$ radians (or $360/n$ degrees) followed by a translation of p/n in the direction of the axis. Again we emphasize that an

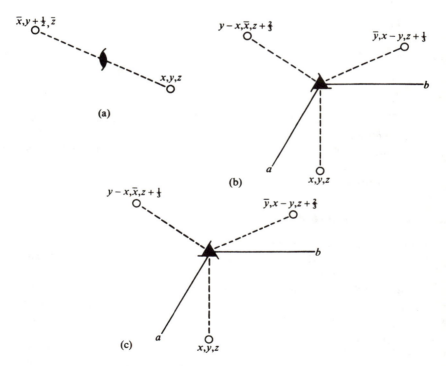

FIG. 4-1 *Screw axes.* (a) 2_1 *axis parallel to* b; (b) 3_1 *axis parallel to* c; (c) 3_2 *axis parallel to* c.

equivalent point is not reached after either the rotation or translation separately, but both motions are part of the total operation. For example, a 2_1 axis involves rotation by 180° followed by translation by one half a unit cell parallel to the axis. For a 3_1 axis the rotation is 120° and the translation is one third of a unit cell, whereas a 3_2 axis implies a rotation of 120° and a translation of two thirds. These cases are illustrated in Fig. 4-1, where the coordinates of equivalent points are shown. In a system referred to hexagonal axes, a 3_1 axis converts the point x,y,z to $\bar{y}, x - y, z + \frac{1}{3}$. Application of 3_1 to $\bar{y}, x - y, z + \frac{1}{3}$ yields $y - x, \bar{x}, z + \frac{2}{3}$, and one more application gives $x,y,z + 1$, which is just one unit cell away from the original point. It should also be apparent from this illustration that the difference between a 3_1 and 3_2 axis is essentially the difference between a right-handed and a left-handed screw. The other possible screw axes are $4_1, 4_2, 4_3, 6_1, 6_2, 6_3, 6_4$, and 6_5.

4-3   Glide planes

The combination of the motions of reflection and translation gives a *glide plane*. The operation consists of reflection in a plane followed by translation. If the glide is parallel to the a axis, the symbol for the glide plane is simply a and the operation is reflection in the plane and translation by $a/2$. For the corresponding cases of glides parallel to b or c, the respective symbols are b and c and these are the three types of axial glide plane. A diagonal glide, denoted by the letter n, involves translations of $(a + b)/2$, $(b + c)/2$, or $(c + a)/2$; that is, the glide direction is parallel to a face diagonal. In the tetragonal, rhombohedral, and cubic systems it is possible to have $(a + b + c)/2$ for the direction of a diagonal glide. Finally, a diamond glide (symbol d) has translations of $(a \pm b)/4$, $(b \pm c)/4$, or $(c \pm a)/4$, or for tetragonal and cubic $(a \pm b \pm c)/4$.

4-4   Space groups

A group whose elements include both the point symmetry elements and the translations of a crystal is called a *space group*. Our study of the point symmetry elements alone led to the important result that there are only thirty-two point groups compatible with lattice translations. To determine the complete list of space groups we should first combine each of the thirty-two point groups with each of

the Bravais lattice types. Thus, point group $2/m$ belongs to the monoclinic system, and the two monoclinic lattices are P and C, so we can
expect $P2/m$ and $C2/m$ as space groups. Our space group symbol will
always consist of a capital letter denoting the centering followed by a
generalization of our Hermann–Mauguin point group symbol to allow
for glide planes and screw axes. In obtaining the space group symbols
in this way, we must remember to include separately such cases as
$Cmm2$, where the twofold axis is perpendicular to the centered face,
and $Amm2$, where the twofold axis is one of the edges of the centered
face. These combinations of point groups with Bravais lattices will give
us a total of seventy-two space groups.

We next have to consider the possibility of replacing each of the
rotations and reflections by the corresponding screw axes and glide
planes. In our example with point group $2/m$, this gives space groups
$P2_1/m$, $P2/c$, $P2_1/c$, and $C2/c$. In this process we must carefully delete
duplications. For example, $P2/a$ is the same as $P2/c$, except for the
naming of the a and c axes. It is less obvious that $C2_1/c$ differs from
$C2/c$ only by a shift of origin, but these are not two different space
groups. Proceeding in this way, we eventually arrive at a list of 230 space
groups, and these are listed in Appendix 1.

4-5   *Relationship between space groups, point groups, and physical properties*

The list of space groups in Appendix 1 has been divided into seven
crystal systems, and each of these has been further divided into point
groups. Thus, associated with the tetragonal point group $4/m$ we have
the six space groups $P4/m$, $P4_2/m$, $P4/n$, $P4_2/n$, $I4/m$, and $I4_1/a$.
Although a structure described by space group $P4/m$ is certainly quite
different from a structure described by space group $P4_2/m$, both crystals
belong to point group $4/m$, and the macroscopic physical properties of
the two crystals will obey the same symmetry conditions. If a physical
property (for example, electrical conductivity) is measured along the
direction $[uvw]$, the property will have the same magnitude along any
of the directions $[\bar{v}uw]$, $[\bar{u}\bar{v}w]$, and $[v\bar{u}w]$, and in the reverse of these
directions. Certain properties may have considerably more symmetry
than this; the electrical conductivity of a tetragonal crystal, for example,
will have all the symmetry of an ellipsoid of revolution.[1] However,

---

[1] See J. F. Nye, *Physical Properties of Crystals*, Oxford Univ. Press, London,
1957, for a treatment of the symmetry of these properties.

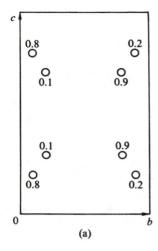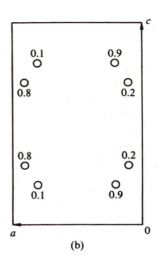

FIG. 4-2   *Arrangement of points in space group P4₂/m. (a) a Axis pointing out from page; origin at lower left. (b) b Axis pointing out from page, a axis to left; origin at lower right.*

all physical properties will have at least this much symmetry, and all crystals belonging to point group $4/m$ have the same relationships between equivalent directions. Figure 4-2 shows a hypothetical structure in space group $P4_2/m$. An atom has coordinates 0.1,0.2,0.3; symmetry operations generate the equivalent points 0.1,0.2,0.7; 0.9,0.8,0.3;   0.9,0.8,0.7;   0.8,0.1,0.8;   0.8,0.1,0.2;   0.2,0.9,0.8; 0.2,0.9,0.2. (These coordinates will be derived in Section 4-6.) The arrangement of atoms is the same whether viewed along the a axis or b axis. That is, Fig. 4-2a is identical with Fig. 4-2b, except for a translation of $\frac{1}{2}c$ (remember that these are periodic structures, so additional points occur in adjacent unit cells). Since the structure looks the same whether viewed along a or b, any physical property will have the same value in these two directions, as long as the physical measurement is not capable of detecting the shift of a few angstrom units corresponding to $\frac{1}{2}c$.

The point group of a crystal may always be obtained from the space group symbol by replacing each screw axis n_p by the proper rotation axis n and each glide plane by a mirror plane m.

4-6   *Equivalent positions*

We have frequently applied the symmetry operations of a group to generate a set of points equivalent to a given point. We now consider a few examples of this process in the case of space groups.

We will use space group $P4_2/m$ as our first illustration of the principles of deriving sets of equivalent positions. Figure 4-3a shows the equivalent points generated by a fourfold rotation axis in the c direction; a counter-clockwise rotation of $90°$ generates a point with coordinates $\bar{y}, x, z$ from a point with coordinates x, y, z. If, for example, our initial point had coordinates $0.1, 0.2, 0.3$, the generated point would be at $-0.2, 0.1, 0.3$ (which may also be written $0.8, 0.1, 0.3$ because of the periodicity of the arrangement). Application of the symmetry operator 4 to point x, y, z, thus gives a point whose x coordinate is $-y$, whose y coordinate is x, and whose z coordinate is z. When 4 is applied to $\bar{y}, x, z$ the result is $\bar{x}, \bar{y}, z$, and still another application gives $y, \bar{x}, z$. One more application would give x, y, z again, so there are just four points related by the operation. The operator required in $P4_2/m$ is 4_2 rather than 4. The difference is simply that a translation of $\frac{1}{2}c$ must be included in each operation, so the equivalent points generated are those in Fig. 4-3b: $x, y, z; \bar{y}, x, \frac{1}{2} + z; \bar{x}, \bar{y}, z; y, \bar{x}, \frac{1}{2} + z$. Space group $P4_2/m$ also includes a reflection perpendicular to the 4_2 axis. For each point x, y, z this

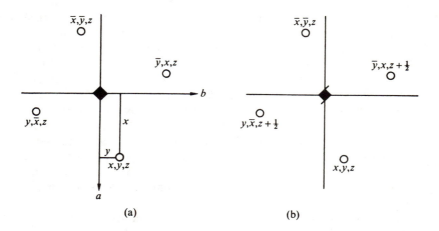

(a)   (b)

FIG. 4-3   *Equivalent points generated by symmetry operations (a)* 4, *(b)* 4_2.

reflection gives a point $x,y,\bar{z}$, so in addition to the four points generated by 4_2 we get four more points by changing the sign of z. Noting that $-(\frac{1}{2}+z) = -\frac{1}{2}-z$ is equivalent to $\frac{1}{2}-z$, we have $x,y,\bar{z}$; $\bar{y},x,\frac{1}{2}-z$; $\bar{x},\bar{y},\bar{z}$; $y,\bar{x},\frac{1}{2}-z$. These eight points constitute the general positions for space group $P4_2/m$, and Fig. 4-2 illustrates these positions for the case where $x = 0.1$, $y = 0.2$, $z = 0.3$; the set of equivalent points within the single unit cell consists of $0.1,0.2,0.3$; $0.8,0.1,0.8$; $0.9,0.8,0.3$; $0.2,0.9,0.8$; $0.1,0.2,0.7$; $0.8,0.1,0.2$; $0.9,0.8,0.7$; $0.2,0.9,0.2$.

4-7   Special positions

Suppose we have a point that is on one of the symmetry elements of the space group, such as $x,y,0$, which lies on the reflection plane of $P4_2/m$. The equivalent points when z is 0 are $x,y,0$; $\bar{y},x,\frac{1}{2}$; $\bar{x},\bar{y},0$; $y,\bar{x},\frac{1}{2}$; $x,y,0$; $\bar{y},x,\frac{1}{2}$; $\bar{x},\bar{y},0$; $y,\bar{x},\frac{1}{2}$. The last four points listed are identical with the first four, so there are only four distinct points in this case. When the number of equivalent points in a set is reduced in this way because the points lie on a symmetry element, the positions are called *special positions*. A complete list of all general and special positions for space group $P4_2/m$ is given in Table 4-1. The first column of this table gives the number or multiplicity of the positions, and the second column gives a notation suggested by Ralph W. G. Wyckoff where the positions

TABLE 4-1   EQUIVALENT POSITIONS OF SPACE GROUP $P4_2/m$

No. of positions	Wyckoff notation	Point symmetry	Positions
8	k	1	x,y,z; $\bar{y},x,\frac{1}{2}+z$; $\bar{x},\bar{y},z$; $y,\bar{x},\frac{1}{2}+z$; $x,y,\bar{z}$; $\bar{y},x,\frac{1}{2}-z$; $\bar{x},\bar{y},\bar{z}$; $y,\bar{x},\frac{1}{2}-z$
4	j	m	$x,y,0$; $\bar{y},x,\frac{1}{2}$; $\bar{x},\bar{y},0$; $y,\bar{x},\frac{1}{2}$
4	i	2	$0,\frac{1}{2},z$; $\frac{1}{2},0,\frac{1}{2}+z$; $0,\frac{1}{2},\bar{z}$; $\frac{1}{2},0,\frac{1}{2}-z$
4	h	2	$\frac{1}{2},\frac{1}{2},z$; $\frac{1}{2},\frac{1}{2},\frac{1}{2}+z$; $\frac{1}{2},\frac{1}{2},\bar{z}$; $\frac{1}{2},\frac{1}{2},\frac{1}{2}-z$
4	g	2	$0,0,z$; $0,0,\frac{1}{2}+z$; $0,0,\bar{z}$; $0,0,\frac{1}{2}-z$
2	f	$\bar{4}$	$\frac{1}{2},\frac{1}{2},\frac{1}{4}$; $\frac{1}{2},\frac{1}{2},\frac{3}{4}$
2	e	$\bar{4}$	$0,0,\frac{1}{4}$; $0,0,\frac{3}{4}$
2	d	$2/m$	$0,\frac{1}{2},\frac{1}{2}$; $\frac{1}{2},0,0$
2	c	$2/m$	$0,\frac{1}{2},0$; $\frac{1}{2},0,\frac{1}{2}$
2	b	$2/m$	$\frac{1}{2},\frac{1}{2},0$; $\frac{1}{2},\frac{1}{2},\frac{1}{2}$
2	a	$2/m$	$0,0,0$; $0,0,\frac{1}{2}$

are labeled consecutively with letters beginning with *a* for the highest symmetry. If we have atoms occupying positions (2*f*) of space group $P4_2/m$, this means that there are two equivalent atoms, at $\frac{1}{2},\frac{1}{2},\frac{1}{4}$ and $\frac{1}{2},\frac{1}{2},\frac{3}{4}$, and the symmetry at these atoms is $\overline{4}$.

4-8 Space group tables in International Tables for X-ray Crystallography

Tables such as Table 4-1 have been derived for all of the space groups and are given in Volume 1 of the *International Tables for X-ray Crystallography*.[2] The practicing crystallographer, therefore, does not always have to derive these tables, although he should be well acquainted with the principles involved in their derivation.

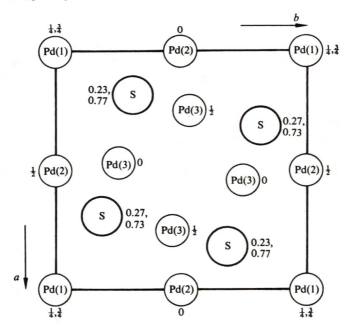

FIG. 4-4 *Structure of PdS projected onto (001). The space group is $P4_2/m$.*

[2] *International Tables for X-Ray Crystallography*, Vol. 1, Kynoch Press, Birmingham, England, 1952. These tables include diagrams, for all except the cubic space groups, which show the locations and orientations of the symmetry elements and the relationships between equivalent positions.

4-9 Examples of the use of space group tables

The use of the space group tables may be clarified by discussions of some actual structures. Palladous sulfide, PdS, has a tetragonal structure with $a = 6.429$ Å and $c = 6.608$ Å. The space group is $P4_2/m$, and there are eight palladium atoms and eight sulfur atoms per unit cell. The sulfur atoms occupy the general positions $(8k)$ with $x = 0.19$, $y = 0.32$, and $z = 0.23$. There are three crystallographically different palladium atoms:

  2 Pd(1) in $2e$
  2 Pd(2) in $2c$
  4 Pd(3) in $4j$ with $x = 0.48$, $y = 0.25$

This information completely describes the structure, and with the aid of Table 4-1 the following positions were derived and used in constructing Fig. 4-4:

$$
\begin{array}{llll}
8 \text{ S at} & 0.19, 0.32, 0.23; & 0.68, 0.19, 0.73 \\
& 0.81, 0.68, 0.23; & 0.32, 0.81, 0.73 \\
& 0.19, 0.32, 0.77; & 0.68, 0.19, 0.27 \\
& 0.81, 0.68, 0.77; & 0.32, 0.81, 0.27
\end{array}
$$

$$
\begin{array}{llll}
2 \text{ Pd(1) at} & 0, 0, \tfrac{1}{4}; & 0, 0, \tfrac{3}{4} \\
2 \text{ Pd(2) at} & 0, \tfrac{1}{2}, 0; & \tfrac{1}{2}, 0, \tfrac{1}{2} \\
4 \text{ Pd(3) at} & 0.48, 0.25, 0; & 0.75, 0.48, \tfrac{1}{2} \\
& 0.52, 0.75, 0; & 0.25, 0.52, \tfrac{1}{2}
\end{array}
$$

Many interesting features of the structure can be obtained from the diagram. For example, each palladium atom is at the center of a distorted square of sulfur atoms, and each sulfur atom is surrounded by a distorted tetrahedron of palladium atoms.

EXERCISE 4-1   Calculate the distances from each type of palladium atom to each of its four sulfur neighbors, and from a sulfur atom to each of its four palladium neighbors, in the PdS structure.

Our second example will be the structure of $HgBr_2$, which has the orthorhombic space group $Bm2_1b$. Note that this symbol does not appear in Appendix 1, but it is equivalent to $Cmc2_1$ on renaming the axes. In $Bm2_1b$, the m stands for a mirror plane normal to a, and this operation converts a point xyz into $\bar{x}yz$. The 2_1 is a screw axis parallel to b, and the point generated is $\bar{x}, \frac{1}{2} + y, \bar{z}$. The b is a glide plane normal to c with a glide component along b, and this produces a point $x, \frac{1}{2} + y, \bar{z}$. The lattice in this example is B centered, which means that for point x, y, z there is an equivalent point $\frac{1}{2} + x, y, \frac{1}{2} + z$. The complete list of positions is given in Table 4-2. The general positions are eightfold, but only four points have been listed. The other four are implied by the heading $(0,0,0; \frac{1}{2},0,\frac{1}{2}) +$, which means that the general positions include the points listed added to $0,0,0$, plus the points listed added to $\frac{1}{2},0,\frac{1}{2}$.

The unit cell dimensions of $HgBr_2$ are $a = 4.624$ Å, $b = 12.445$ Å, $c = 6.798$ Å, and there are four Hg atoms and eight Br atoms per unit cell. The Hg atoms, therefore, must lie on mirror planes, and they occupy the special positions $(4a)$ with $y = 0.000$, $z = 0.334$. Rather than being in general positions, the eight Br atoms occupy two sets of $(4a)$ positions: Br(1) has $y = 0.132$, $z = 0.056$, and Br(2) has $y = 0.368$, $z = 0.389$. The structure is shown in Fig. 4-5.

EXERCISE 4-2   Describe the surroundings of a mercury atom in the $HgBr_2$ structure. Calculate the distances from a mercury atom to the two bromine atoms nearest to it, and calculate the Br—Hg—Br angle formed by these atoms.

EXERCISE 4-3   Derive the general positions for space group $Pmc2_1$. Obtain the special positions by considering the coordinates of points lying on the mirror plane.

As our next example, we compare space groups $P321$ and $P3_121$. In the trigonal system the threefold axis is along c. The twofold axes in these cases are parallel with the a and b axes of the hexagonal cell. The 1 in each of these symbols serves to distinguish these space groups from $P312$ and $P3_112$, where the twofold axes are normal to a and b. The

TABLE 4-2   EQUIVALENT POSITIONS OF SPACE GROUP $Bm2_1b$

No. of positions	Wyckoff notation	Point symmetry	Positions $(0,0,0; \frac{1}{2},0,\frac{1}{2}) +$
8	b	1	$x,y,z; \bar{x},y,z; \bar{x},\frac{1}{2} + y,\bar{z}; x,\frac{1}{2} + y,\bar{z}$
4	a	m	$0,y,z; 0,\frac{1}{2} + y,\bar{z}$

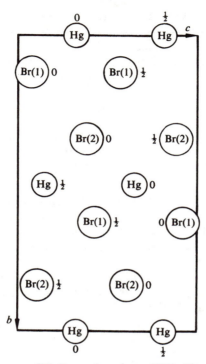

FIG. 4-5 *Structure of HgBr₂ projected onto (100). The space group is Bm2₁b.*

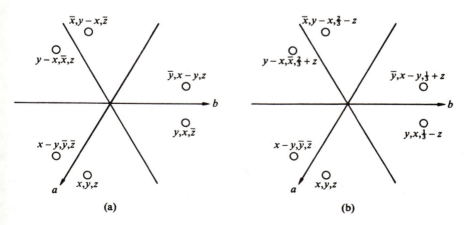

FIG. 4-6 *Equivalent points in (a) P321, (b) P3₁21.*

TABLE 4-3 EQUIVALENT POSITIONS OF SPACE GROUP $P3_121$

No. of positions	Wyckoff notation	Point symmetry	Positions
6	c	1	$x,y,z; \ y,x,\frac{1}{3}-z; \ \bar{y},x-y,\frac{1}{3}+z;$ $\bar{x},y-x,\frac{2}{3}-z; \ y-x,\bar{x},\frac{2}{3}+z;$ $x-y,\bar{y},\bar{z},$
3	b	2	$x,0,\frac{1}{2}; \ 0,x,\frac{5}{6}; \ \bar{x},\bar{x},\frac{1}{6}$
3	a	2	$x,0,0; \ 0,x,\frac{1}{3}; \ \bar{x},\bar{x},\frac{2}{3}$

derivation of the equivalent positions is straightforward. The threefold axis applied to x,y,z generates a point with hexagonal coordinates $\bar{y},x-y,z$, and if the axis is 3_1 instead of 3 the equivalent point is $\bar{y},x-y,\frac{1}{3}+z$. The twofold axis along a converts x,y,z into $x-y,\bar{y},\bar{z}$. These points are shown in Fig. 4-6.

The complete set of equivalent points for $P3_121$ is given in Table 4-3. (These differ from the set given in the *International Tables for X-ray Crystallography* by a rotation of 60°.)

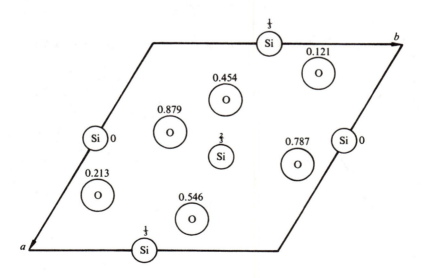

FIG. 4-7 *Structure of α-quartz. The space group is $P3_121$.*

The SiO_2 (quartz) structure in Fig. 4-7 provides an example of this space group. The unit cell has dimensions $a = 4.913$ Å, $c = 5.405$ Å, and contains three SiO_2. The Si atoms occupy positions $(3a)$ with $x = 0.465$; the O atoms occupy the general positions $(6c)$ with $x = 0.272$, $y = 0.420$, $z = 0.454$. Each silicon atom is surrounded by a nearly regular tetrahedron of oxygen atoms: the neighbors of the Si at $0.535, 0.535, \frac{2}{3}$ are the oxygen atoms at $0.272, 0.420, 0.454$; $0.420, 0.272, 0.879$; $0.852, 0.580, 0.546$; and $0.580, 0.852, 0.787$.

EXERCISE 4-4   Derive the general positions for $P312$. Also obtain the special positions for points along the threefold axis that passes through the origin. Are there any threefold axes in this space group that do not pass through the origin?

EXERCISE 4-5   Calculate the distance from a silicon atom to each of its four oxygen neighbors in the α-quartz structure.

4-10   Equivalent positions and the choice of origin

The general positions of the orthorhombic space group $Bmab$ ($Cmca$ with the axes renamed) may be easily derived.

$$(0,0,0; \tfrac{1}{2}, 0, \tfrac{1}{2}) +$$

$$
\begin{array}{ll}
x, y, z; & \bar{x}, y, z \\
\tfrac{1}{2} + x, \bar{y}, z; & \tfrac{1}{2} - x, \bar{y}, z \\
x, \tfrac{1}{2} + y, \bar{z}; & \bar{x}, \tfrac{1}{2} + y, \bar{z} \\
\tfrac{1}{2} + x, \tfrac{1}{2} - y, \bar{z}; & \tfrac{1}{2} - x, \tfrac{1}{2} - y, \bar{z}
\end{array}
$$

The point group associated with this space group is mmm, which is centrosymmetric; that is, a center of symmetry is one of its symmetry elements. However, these equivalent positions do not include $\bar{x}, \bar{y}, \bar{z}$, which is characteristic of a center of symmetry, and a centrosymmetric point group always implies a centrosymmetric space group. A structure based on space group $Bmab$ is, indeed, centrosymmetric, but the center of symmetry is not at the origin $0, 0, 0$, which was defined by the intersection of the m, a, and b planes. The point $\tfrac{1}{4}, \tfrac{1}{4}, 0$ is a center of symmetry in this case, and, if we choose this point as our origin, our set of positions will include $\bar{x}, \bar{y}, \bar{z}$. There are advantages in having the origin at a center of symmetry, one of which will become apparent when we discuss X-ray diffraction. We, therefore, shift the origin by subtracting $\frac{1}{4}, \frac{1}{4}, 0$ from each of the above points. We then have

$$(0,0,0; \tfrac{1}{2},0,\tfrac{1}{2}) +$$

$$
\begin{array}{ll}
\tfrac{3}{4}+x, \tfrac{3}{4}+y, z; & \tfrac{3}{4}-x, \tfrac{3}{4}+y, z \\
\tfrac{1}{4}+x, \tfrac{3}{4}-y, z; & \tfrac{1}{4}-x, \tfrac{3}{4}-y, z \\
\tfrac{3}{4}+x, \tfrac{1}{4}+y, \bar{z}; & \tfrac{3}{4}-x, \tfrac{1}{4}+y, \bar{z} \\
\tfrac{1}{4}+x, \tfrac{1}{4}-y, \bar{z}; & \tfrac{1}{4}-x, \tfrac{1}{4}-y, \bar{z}
\end{array}
$$

We now let $x' = \tfrac{3}{4} + x, y' = \tfrac{3}{4} + y, z' = z$, or $x = x' - \tfrac{3}{4}, y = y' - \tfrac{3}{4}, z = z'$, to obtain

$$
\begin{array}{ll}
x', y', z'; & \tfrac{1}{2}-x', y', z' \\
\tfrac{1}{2}+x', \tfrac{1}{2}-y', z'; & \bar{x}', \tfrac{1}{2}-y', z' \\
x', \tfrac{1}{2}+y', \bar{z}' & \tfrac{1}{2}-x', \tfrac{1}{2}+y', \bar{z}' \\
\tfrac{1}{2}+x', \bar{y}', \bar{z}'; & \bar{x}', \bar{y}', \bar{z}'
\end{array}
$$

and when we drop the primes we have the list given in Table 4-4.

Solid chlorine at $-160°C$ has a structure based on this space group. The unit cell dimensions are $a = 6.24$ Å, $b = 8.26$ Å, $c = 4.48$ Å, and there are eight atoms in the positions $(8f)$ with $y = 0.100$, $z = 0.370$. The structure is shown in Fig. 4-8. The atoms are joined in diatomic molecules; for example, the atom at $\frac{1}{4}, 0.40, 0.87$ is bonded to the atom at $\frac{1}{4}, 0.60, 0.63$.

TABLE 4-4 EQUIVALENT POSITIONS OF SPACE GROUP *Bmab*. ORIGIN AT CENTER ($\bar{1}$)

No. of positions	Wyckoff notation	Point symmetry	Positions
			$(0,0,0; \tfrac{1}{2},0,\tfrac{1}{2}) +$
16	*g*	1	$x,y,z; \tfrac{1}{2}-x,y,z; \tfrac{1}{2}+x,\tfrac{1}{2}-y,z;$
			$\bar{x},\tfrac{1}{2}-y,z; x,\tfrac{1}{2}+y,\bar{z}; \tfrac{1}{2}-x,\tfrac{1}{2}+y,\bar{z};$
			$\tfrac{1}{2}+x,\bar{y},\bar{z}; \bar{x},\bar{y},\bar{z}$
8	*f*	*m*	$\tfrac{1}{4},y,z; \tfrac{3}{4},\tfrac{1}{2}-y,z; \tfrac{1}{4},\tfrac{1}{2}+y,\bar{z}; \tfrac{3}{4},\bar{y},\bar{z}$
8	*e*	2	$0,\tfrac{1}{4},z; \tfrac{1}{2},\tfrac{1}{4},z; 0,\tfrac{3}{4},\bar{z}; \tfrac{1}{2},\tfrac{3}{4},\bar{z}$
8	*d*	2	$x,0,\tfrac{1}{4}; \tfrac{1}{2}-x,0,\tfrac{1}{4}; x,\tfrac{1}{2},\tfrac{3}{4}; \tfrac{1}{2}-x,\tfrac{1}{2},\tfrac{3}{4}$
8	*c*	$\bar{1}$	$0,0,0; \tfrac{1}{2},0,0; \tfrac{1}{2},\tfrac{1}{2},0; 0,\tfrac{1}{2},0$
4	*b*	2/*m*	$\tfrac{1}{4},0,\tfrac{3}{4}; \tfrac{3}{4},\tfrac{1}{2},\tfrac{3}{4}$
4	*a*	2/*m*	$\tfrac{1}{4},0,\tfrac{1}{4}; \tfrac{3}{4},\tfrac{1}{2},\tfrac{1}{4}$

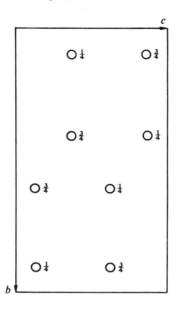

FIG. 4-8   *Structure of Cl_2 at $-160°C$. The space group is Bmab.*

EXERCISE 4-6   Calculate the bond length in the Cl_2 molecule at $-160°C$. Also, calculate the distance from a chlorine atom to some of the chlorine atoms in adjacent molecules.

EXERCISE 4-7   Tridymite, the high-temperature form of SiO_2, is hexagonal with $a = 5.03$, $c = 8.22$ Å. The space group is $P6_3/mmc$, and there are four SiO_2 units per unit cell. The Si atoms occupy the positions $\frac{1}{3},\frac{2}{3},0.44$; $\frac{2}{3},\frac{1}{3},0.56$; $\frac{2}{3},\frac{1}{3},0.94$; $\frac{1}{3},\frac{2}{3},0.06$. Two oxygen atoms, O(1), are in the positions $\frac{1}{3},\frac{2}{3},\frac{1}{4}$; $\frac{2}{3},\frac{1}{3},\frac{3}{4}$. The other six oxygens O(2) are in the positions $\frac{1}{2},0,0$; $0,\frac{1}{2},0$; $\frac{1}{2},\frac{1}{2},0$; $\frac{1}{2},0,\frac{1}{2}$; $0,\frac{1}{2},\frac{1}{2}$; $\frac{1}{2},\frac{1}{2},\frac{1}{2}$.

(a) Draw a diagram of this structure projected onto the (001) plane. Show all the atoms in at least one unit cell. Label each atom with the atomic type [Si, O(1) or O(2)] and with the z coordinate.

(b) From your diagram deduce the point symmetry at Si, at O (1) and at O (2).

(c) Calculate the distance from Si to each of the two types of oxygen atom to which it is bonded.

EXERCISE 4-8   Derive the general positions corresponding to the space group symbols $C2/c$ and $C2_1/c$ and show that these refer to the same space group, differing only in the choice of origin.

Chapter 5

X - RAY DIFFRACTION

We are now ready to study the diffraction of X rays by crystals. In this chapter we will discuss the determination of unit cell geometry, which involves using X rays to obtain the unit cell dimensions, the lattice type, the crystal system, and the possible space groups. We will find it necessary to know how the intensities of the diffracted X-ray beams depend upon the locations of the atoms within a unit cell, and this will lead us into a treatment of Fourier series and structure factors. Methods for obtaining the atomic positions from measured values of the intensities will be treated in Chapter 6.

5-1   Periodicity and structural information

A simple analogy may illustrate why crystallographic studies are so powerful in determining molecular structure. We first consider several identical molecules with random orientations. We suppose that we have available some instrument that is capable of supplying structural information. For example, our experimental method may involve interaction of the molecules with electromagnetic radiation, and the pattern of scattered radiation provides the clues from which we must determine the structure. Since a single interaction of a photon with a single molecule cannot provide us with sufficient data to deduce the structure, we must repeat the process many times with many molecules. We will finally obtain a description or picture of the molecular structure, but our picture will be averaged over all of the orientations of the molecules and it will not resemble a snapshot of a single molecule. Such a picture might be adequate for deducing the structure of a simple molecule, but if the molecule is at all complex the interpretation may be very difficult and uncertain.

We now carry out the same experiment on a crystalline arrangement of molecules. In this case the superposition of the pictures of many molecules looks like a single molecule, and even if the molecule is complex we will know its structure unambiguously.

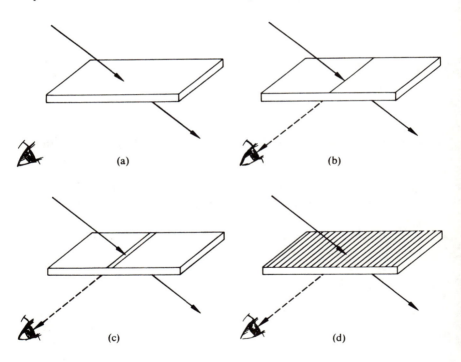

FIG. 5-1   (a) *Light passes through clear glass. None scattered to observer.* (b) *A scratch on the glass scatters light to observer.* (c) *Radiation scattered by two scratches experiences interference.* (d) *A diffraction grating. Scattered radiation observed only at certain angles.*

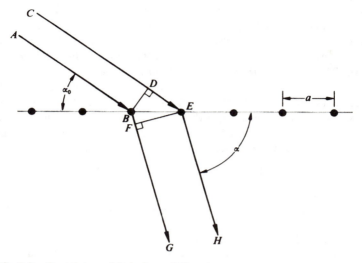

FIG. 5-2 *Scattering of light by a diffraction grating with repeat distance a.*

Our actual experimental method cannot, of course, consist of photographing individual molecules, but the point of this analogy is that information is lost when the molecules have random orientations, whereas a crystalline arrangement can provide sufficient information for deducing the structure.

5-2 The diffraction grating

Diffraction effects with visible light were first observed more than 300 years ago, and the diffraction grating was invented by Fraunhofer in 1820. The geometric principles are demonstrated in Fig. 5-1. A beam of light passes through the clear sheet of glass in Fig. 5-1a. In Fig. 5-1b there is a scratch on the glass which scatters light; the observer sees a scratch. The two parallel scratches in Fig. 5-1c both scatter light, but interference can occur between the two scattered rays, and the intensity will depend upon the angle between the incident ray and the line of observation. Each of the scratches on the slide in Fig. 5-1d will scatter light, but the mutual interference of these scattered rays makes the observed intensity practically zero except near certain angles. The derivation of the angles at which the scattered intensity is maximum is based on Fig. 5-2. The incident beam makes angle α_0 with the diffraction grating. The incident ray CE travels farther than AB before reaching the grating, and the scattered ray BG travels farther than EH after passing the grating. The difference in path lengths of the beams $CDEH$ and $ABFG$ is $DE - BF$, and this difference must be equal to a whole number of wavelengths if the high intensity characteristic of constructive interference is to be observed at angle α. Therefore, $DE - BF = n\lambda$, where λ is the wavelength of the light and n is an integer (n can be any positive or negative integer, $-2,-1,1,2,...$). By simple geometry $DE = a\cos\alpha_0$ and $BF = a\cos\alpha$, where a is the repeat distance, so

$$a(\cos\alpha_0 - \cos\alpha) = n\lambda \tag{5-1}$$

This is the linear diffraction grating formula. For a given α_0, α depends upon the repeat distance, the wavelength, and the integer n.

EXERCISE 5-1 Light of wavelength 5000 Å is incident at 60° to a grating with 4000 lines per centimeter. Calculate the angles at which the first- and second-order maxima will occur. (That is, evaluate α for $n = 1$ and $n = 2$.)

5-3   Diffraction of X rays by crystals

X rays were discovered by Wilhelm Röntgen in 1895. In 1912 it was still not known whether X rays consisted of particles or whether they were electromagnetic waves. It was known that if the wave hypothesis were correct, the wavelengths must be of the order of 1 Å (10^{-8} cm). It was believed impossible to use a grating to measure such short wavelengths, since all grating experiments that had been performed had involved wavelengths of the same order of magnitude as the grating spacing. (Gratings actually are now the source of our most accurate measurements of X-ray wavelengths.)

EXERCISE 5-2 A grating with 4000 lines per centimeter is used to study X rays of wavelength 1.00 Å. Calculate the deviation of the beam (i.e., $\alpha - \alpha_0$) for first-order maxima when α_0 is 10°, 1°, and 0.1°. Do the results suggest how measurements of X-ray wavelengths using gratings may be carried out?

In 1912 there was also no direct evidence for the structure of crystals, although there were reasons for believing that crystals had periodic arrangements of atoms with interatomic distances of the order of 1 Å. Max von Laue, at the University of Munich in Germany, suggested that the periodic structure of a crystal might be used to diffract X rays just as gratings are used to produce diffraction patterns with visible light. This proposal was based on three assumptions: (1) crystals are periodic, (2) X rays are waves, (3) the X-ray wavelength is of the same order of magnitude as the repeat distance in crystals. Friedrich and Knipping carried out an experimental test of von Laue's suggestion by irradiating a crystal of $CuSO_4 \cdot 5H_2O$ with X rays. The detection of diffraction confirmed von Laue's suggestion and launched the science of X-ray crystallography.[1]

5-4   *The Laue equations*

The observation of diffracted X rays only in certain allowed directions is entirely analogous to the diffraction of light by a grating. In both the crystal and the grating, the allowed angles are determined only by the repeat distance of the periodic structure and the wavelength of the radiation. The detailed structure of the rulings on the grating will affect the intensities at the allowed angles, but the grating spacing is the only grating property included in Eq. (5-1). Similarly, the distances between identical points in a crystal will comprise the only information required for the corresponding crystallographic equations. Since crystals are periodic in three dimensions, three equations are required.

$$a(\cos \alpha_0 - \cos \alpha) = h\lambda \qquad (5\text{-}2a)$$

$$b(\cos \beta_0 - \cos \beta) = k\lambda \qquad (5\text{-}2b)$$

$$c(\cos \gamma_0 - \cos \gamma) = l\lambda \qquad (5\text{-}2c)$$

These are called the Laue equations. (Do not confuse these angles with the unit cell angles, for which the same symbols were used.) The angles between the incident X-ray beam and the unit cell axes a, b, c are α_0, β_0, and γ_0, and α, β, and γ are the corresponding angles for the diffracted beam. Constructive interference will occur only for values of these six angles for which h, k, and l in Eqs. (5-2) are integers.

---

[1] An account of the history of X-ray crystallography, including a reproduction of Friedrich and Knipping's first successful photograph, may be found in *Fifty Years of X-ray Diffraction* (P. P. Ewald, ed.), published for the International Union of Crystallography by N. V. A. Oosthoek's Uitgeversmaatschappij, Utrecht, 1962.

5-5   Rotating crystal method

The Laue equations (5-2) may be applied directly in interpreting the geometry of X-ray diffraction. One such application is in the rotating crystal method, illustrated in Fig. 5-3. A crystal is rotated continuously about one of the unit cell axes, which we will call a, and the incident X-ray beam is normal to this axis. The angle α_0 is, therefore, 90°, and $\cos \alpha_0 = 0$. If $h = 0$, Eq. (5-2a) is satisfied if $\alpha = 90°$. There will be various allowed directions for the diffracted beam for the case $h = 0$, corresponding to the solutions of Eqs. (5-2b) and (5-2c), but these directions will all lie in the plane normal to the rotation axis a. As the crystal rotates about a, Eq. (5-2a) will always be satisfied, and orientations will be reached at which Eqs. (5-2b) and (5-2c) are also satisfied. Orientations that simultaneously satisfy all three Laue equations will be achieved more often than might be expected because the six angles

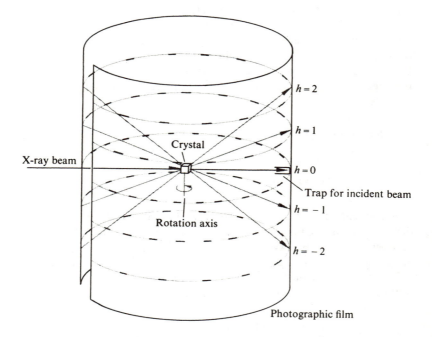

FIG. 5-3   *Rotating crystal method. X-ray beam perpendicular to a axis of crystal. The allowed directions form a cone for each value of h.*

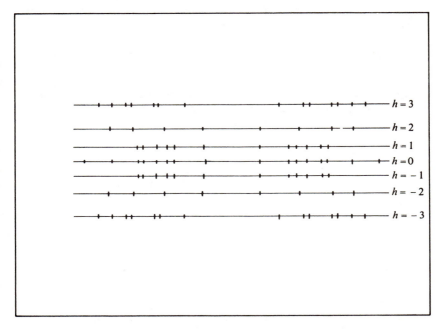

FIG. 5-4 *Rotation pattern obtained by unrolling and developing the film shown in Fig. 5-3.*

$\alpha_0, \beta_0, \gamma_0, \alpha, \beta, \gamma$ are not all independent. (If, for example, a, b, and c are mutually perpendicular, $\cos^2\alpha_0 + \cos^2\beta_0 + \cos^2\gamma_0 = 1$ and $\cos^2\alpha + \cos^2\beta + \cos^2\gamma = 1$). As the crystal rotates through such a position, a ray of diffracted radiation is sent out in the appropriate direction.

For each value of h other than 0, there is a cone of diffracted radiation. The half-angle of this cone is the complement of the angle α of Eq. (5-2a). The diffracted radiation may be intercepted by a photographic film, and if the film is wrapped around the crystal as a cylinder (Fig. 5-3), the unwrapped and developed film will show series of spots on straight lines, as shown in Fig. 5-4. Each straight row corresponds to one value of h. The length of the crystal axis a may be obtained from the distances between these straight rows by the formula

$$a = \frac{h\lambda}{\sin\tan^{-1}(y/r)} \qquad (5\text{-}3)$$

where λ is the wavelength of the X rays, r is the radius of the cylindrical film, and y is the distance on the film from row 0 to row h.

EXERCISE 5-3   Derive Eq. (5-3) from Eqs. (5-2).

EXERCISE 5-4   A rotation photograph was taken with X rays of wavelength 1.542 Å and a film diameter of 57.3 mm. A millimeter scale placed next to the developed film gave the following readings:

$h = 3$	5.40 mm
$h = 2$	22.44 mm
$h = 1$	31.83 mm
$h = 0$	39.40 mm
$h = -1$	46.96 mm
$h = -2$	56.35 mm
$h = -3$	73.20 mm

Calculate the value of a for each row, and average the results.

Although one unit cell dimension is easily obtained from a rotation pattern, it is not as simple to determine the other two lengths and the three angles. The lengths could be obtained by remounting the crystal about each of two other axes in turn, but this would be tedious and time consuming. Some simplification may be achieved by means of oscillation photographs, which are taken with the crystal oscillating through a limited angular range instead of undergoing complete rotation. This procedure facilitates deducing the integers k and l to be assigned to each spot on the row characterized by h. The problem is complicated by having two unknown lengths and four unknown angles in Eqs. (5-2b) and (5-2c), but all necessary information can be acquired by carefully correlating the appearance of the spots with the angle of rotation.

5-6   *Bragg's law*

Shortly after the discovery of X-ray diffraction, W. H. Bragg and his son, W. L. Bragg, discovered that the geometry of the process is analogous to the reflection of light by a plane mirror. As was discussed in Section 3-8, a consequence of the three-dimensional periodicity of a crystal structure is that perpendicular to certain directions it is possible to construct sets of many planes that are parallel with each other, equally spaced, and contain identical atomic arrangements.[2] If an incident X-ray beam makes an angle θ with such a set of planes, the "reflected" beam also makes an angle θ with the planes, as in the case of optical reflection. It, of course, follows that the angle between the incident and reflected rays is 2θ.

Physically, the process consists of the scattering of X rays by the electron clouds surrounding the atoms of the crystal. The observed pattern is the result of the constructive and destructive interference of the radiation scattered by all of the atoms, and the analogy to ordinary reflection is a result of the regularity of the atomic arrangement in a crystal.

Since there are large numbers of parallel planes involved in scattering X rays, reflections from successive planes will interfere with each other, and there will be constructive interference only when the difference in path length between rays from successive planes is equal to a whole number of wavelengths. This is illustrated in Fig. 5-5 where X rays of wavelength λ are incident at angle θ on a set of planes with spacing d. The ray striking the second plane travels a distance $AB + BC$ farther than the ray striking the first plane. These two rays will be in phase only

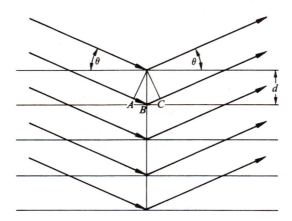

FIG. 5-5 *An X-ray beam makes angle θ with a set of planes with interplanar spacing d. For constructive interference $n\lambda = 2d\sin\theta$.*

[2] Review Sections 3-8, 3-9, and 3-10 concerning planes in crystals.

if

$$AB + BC = n\lambda$$

where n is some integer. From elementary geometry

$$AB = BC = d \sin \theta$$

Therefore,

$$2d \sin \theta = n\lambda \qquad (5\text{-}4)$$

and this is the well-known Bragg's law. Equation (5-4) provides no information other than that given by the Laue equations, but the interpretation of X-ray diffraction patterns is frequently easier in terms of Bragg's law since only one measured angle is required.

5-7   Generalization of Miller indices

In applying Bragg's law to the interpretation of X-ray diffraction patterns, it will be advantageous to discard the restriction of Section 3-9 that the three Miller indices of a plane have no common divisor. Suppose we use X rays of wavelength 1.542 Å to record the (100) reflection from a set of planes that have a d-spacing of 4.00 Å. According to Eq. (5-4), $\sin \theta = (1 \times 1.542)/(2 \times 4.00) = 0.193$ and $\theta = 11.15°$ for this reflection. But this is only the first-order reflection, and equally valid solutions of Eq. (5-4) result from using $n = 2$, $n = 3$, and so on. For $n = 2$, we would have $\sin \theta = 0.386$ or $\theta = 22.7°$, and this second-order reflection can also be observed. If, on the other hand, we assume the existence of a set of (200) planes, the d spacing for (200) is 2.00 Å, and planes with this spacing would give a first-order reflection with $\sin \theta = (1 \times 1.542)/(2 \times 2.00) = 0.386$ or $\theta = 22.7°$. Thus, as far as X-ray diffraction is concerned, there is no distinction between the second-order reflection from (100) and the first-order reflection from (200). It is convenient to avoid referring to different orders of reflection and merely absorb the factor n of Eq. (5-4) in the Miller indices. This is the procedure we will follow, even though it is illogical from the standpoint of classical crystallography. [After all, the (200) planes in a primitive unit cell do not pass through points equivalent to those on (100) planes.] It will be convenient for us, therefore, to always use Bragg's law in the form

$$2d \sin \theta = \lambda \qquad (5\text{-}5)$$

and the d spacing will be calculated by Eq. (3-3) regardless of whether or not the indices are relatively prime.

5-8   *Weissenberg camera*

There are various experimental techniques for accumulating X-ray diffraction data, but we will give brief descriptions of only two photographic methods, besides the rotation camera. More complete information on these methods, on other types of cameras, and on counter techniques can be found in the numerous advanced textbooks of crystallography, and a visit to an X-ray diffraction laboratory will contribute more to understanding these instruments than any amount of explanation.

The Weissenberg camera is essentially a rotation camera with a cylindrical metal screen that allows only one layer line at a time to reach the film. For example, a zero-layer Weissenberg photograph may be taken with the arrangement of Fig. 5-3 by allowing only the radiation of the $h = 0$ spots to reach the film. A first-layer Weissenberg photograph would have just the $h = 1$ reflections, and so forth. If this were all there were to it, the photograph would consist only of one of the straight rows of spots of Fig. 5-4. However, the film is translated back and forth in synchronization with the rotation (see Fig. 5-6), so that the spots no longer lie on the straight line but are distributed over the whole film in a manner such that the vertical coordinate in Fig. 5-4 depends upon the rotational coordinate of the crystal. A typical Weissenberg application might rotate the crystal through 200° while

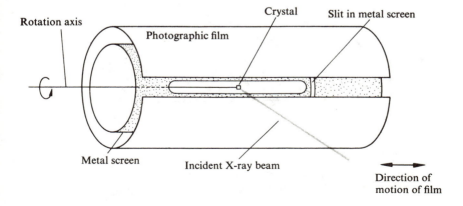

FIG. 5-6   *Principle of the Weissenberg camera.*

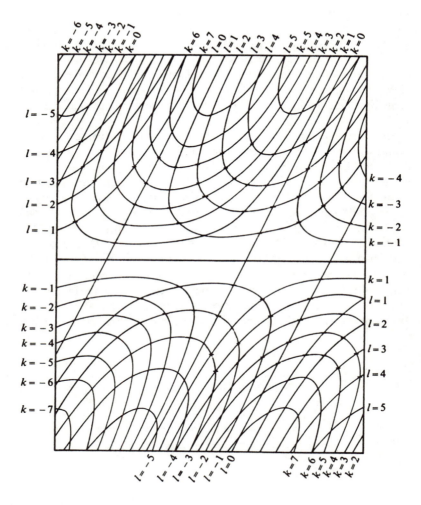

FIG. 5-7   *Appearance of a Weissenberg photograph. With orientation about a, all reflections on the film have the same value of h (0 for a zero layer). Curves connecting spots with constant k or constant l simplify indexing.*

the film is translated parallel to the rotation axis by 100 mm. The direction of rotation is then reversed and the crystal is rotated 200° in the reverse direction while the film is translated back 100 mm. The cycle is repeated many times until a sufficiently intense exposure has been attained. The Weissenberg camera tremendously simplifies the interpretation of X-ray diffraction patterns. With its aid the experienced crystallographer can quickly deduce the unit cell size and shape, he can determine the crystal system and some possible space groups, and he can assign correct Miller indices to all of the reflections.[3] The typical appearance of a Weissenberg photograph is shown in Fig. 5-7, which illustrates a zero layer with orientation about a; curves corresponding to series of reflections with constant k or constant l have been drawn in, and these curves make rapid indexing of the spots possible.

5-9   Buerger precession camera

In the Buerger precession camera, a crystal axis makes a constant angle with the X-ray beam; a typical case might use an angle of 30°. The crystal precesses so as to maintain this angle, and during this precessing motion various planes successively pass through reflecting position. A suitable arrangement with a moving screen and film is used to record

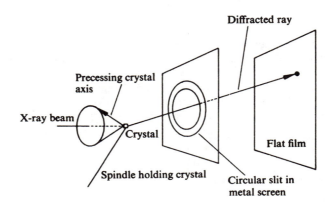

FIG. 5-8   *Principle of the precession camera.*

[3] The experimental details of the Weissenberg camera are given in M. J. Buerger, *X-ray Crystallography*, Wiley, New York, 1942.

only one type of reflection at a time (see Fig. 5-8). For example, if the precessing axis is b, a zero-layer precession photograph contains only $h0l$ reflections, a first layer contains $h1l$, and so on.[4]

5-10   Comparison of Weissenberg and precession techniques

The Weissenberg camera and the precession camera each has its advantages. The cameras are best used to complement each other since they provide different information with the same crystal orientation. If, for example, a crystal is oriented about the a axis, a zero-layer Weissenberg photograph gives $0kl$ reflections. Without remounting the crystal $h0l$ reflections can be obtained with a precession camera by letting the b axis precess about the X-ray beam, and $hk0$ reflections can be recorded by letting the c axis be the precessing axis. A first layer of each type will give $1kl$, $h1l$, and $hk1$ reflections which will usually suffice for all except a complete structure determination of a complicated crystal in three dimensions, and a few upper layers of each type will usually give all possible reflections.[5]

EXERCISE 5-5   Calculate the θ values at which the following reflections would appear from a monoclinic crystal with $a = 5.50$ Å, $b = 8.05$ Å, $c = 7.68$ Å, $\beta = 110.0°$ (assume an X-ray wavelength of 1.542 Å): (a) 010; (b) 020; (c) 321; (d) 32$\bar{1}$.

EXERCISE 5-6   Crystals of sodium thiosulfate, $Na_2S_2O_3$, are monoclinic with four molecules per unit cell. A crystal was oriented about the b axis, and a rotation pattern taken with a 57.3-mm diameter camera and radiation with wavelength 0.7107 Å had a distance of 5.02 mm between the $k = 1$ and $k = -1$ rows. The following θ values were obtained from a zero-layer Weissenberg photograph, which was also taken with 0.7107 Å X rays:

(200)	6.30°
(201)	6.94°
(20$\bar{1}$)	6.55°

(a) Calculate the length of b.

---

[4] For more details, see M. J. Buerger, *The Precession Method in X-ray Crystallography*, Wiley, New York, 1964.

[5] An account of the instrumentation used in X-ray diffraction studies is given in six articles by R. Rudman, *J. Chem. Educ.* **44**, A7, A99, A187, A289, A399, A499 (1967).

(b) Calculate the lengths of a and c and the magnitude of the angle β. (Note: β could have been determined directly from the Weissenberg photograph with greater accuracy than this calculation gives.)

(c) Calculate the volume of the unit cell and the density of anhydrous sodium thiosulfate.

EXERCISE 5-7   Calculate the angles θ at which the following reflections would appear from a hexagonal crystal with $a = 6.50$ Å, $c = 9.90$ Å (assume an X-ray wavelength of 1.658 Å):

h	k	l
1	0	2
0	1	2
1	$\bar{1}$	2
1	1	2
1	0	$\bar{2}$

5-11   Information obtained from diffraction patterns

Our development so far has indicated how the dimensions of a unit cell may be determined from measurements of the positions of the diffraction spots on a photographic film. In order to deduce the crystal system, the symmetry of the diffraction pattern must be determined, and the relationship of the diffraction symmetry to the symmetry of the crystal structure will be studied in Section 5-21. In Sections 5-22, 5-23, and 5-24, we will encounter systematic absences of certain types of reflections that supply evidence of the presence of nonprimitive unit cells, of glide planes, and of screw axes, and this information will aid in deducing possible space groups. In Chapter 6, we will investigate methods of determining the locations of the atoms within a unit cell. Our approach to all of these topics requires that we consider the dependence of the intensities of X-ray reflections on the atomic positions. A thorough treatment of this subject would require extensive studies of the interaction of X rays with single electrons, of the interference phenomena occurring when X rays are scattered by the electrons in atoms, and of the interference between rays scattered by different atoms. Although we will gloss over most of the complicated theory and try to achieve understanding in terms of elementary concepts, the mathematics of the next few sections cannot be avoided.

5-12   *Electron density function*

Since X rays are scattered by the electrons of the atoms, a crystal property suitable for our mathematical treatment is the electron density. We will use $\rho(xyz)$ to represent the electron density or number of electrons per unit volume near the point in the unit cell that has coordinates x, y, z.

The electron density is a periodic function. A mathematical statement of the periodicity is

$$\rho(x + p, y + q, z + r) = \rho(xyz) \tag{5-6}$$

where p, q, and r are any integers.

5-13   *Fourier series*

It is frequently useful to be able to express a function by means of a Fourier series; that is, as a sum of sine and cosine terms with appropriate coefficients. Fourier expansion is particularly advantageous when the function is periodic, and the Fourier expansion of the electron density will be the keystone of our development.[6]

As an example of a Fourier expansion, we consider the periodic function in Fig. 5-9. The function $f(x)$ is $+1$ when x is between 0 and 1; -1 when x is between 1 and 2; $+1$ when x is between 2 and 3; and so on.

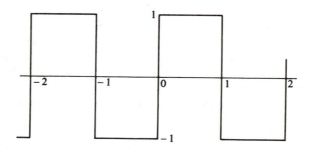

FIG. 5-9   *A periodic function of amplitude 1 and period 2.*

---

[6] Discussions of Fourier series are included in most textbooks of advanced calculus. The results of interest to crystallographers are given in *International Tables for X-ray Crystallography*, Vol. II, Kynoch Press, Birmingham, England, 1959.

We write

$$f(x) = \sum_n A_n \cos n\pi x + \sum_n B_n \sin n\pi x \qquad (5\text{-}7)$$

where A_n and B_n are coefficients that must be determined, and the summation is over all positive and negative integers n. An equation completely equivalent to Eq. (5-7) is

$$f(x) = \sum_n C_n e^{-in\pi x} \qquad (5\text{-}8)$$

where $i = \sqrt{-1}$. These equations are equivalent because the complex exponential term is related to the trigonometric functions by

$$e^{-in\pi x} = \cos n\pi x - i \sin n\pi x \qquad (5\text{-}9)$$

which may be verified by Taylor series expansions of both sides. We will use Eq. (5-9) repeatedly in our subsequent discussion, and you should be completely familiar with it. Two important corollaries of Eq. (5-9) are

$$e^{in\pi x} + e^{-in\pi x} = 2\cos n\pi x \qquad (5\text{-}10)$$

$$e^{in\pi x} - e^{-in\pi x} = 2i\sin n\pi x \qquad (5\text{-}11)$$

In order to evaluate the coefficients C_n in Eq. (5-8) for the periodic function of Fig. 5-9, we multiply both sides of Eq. (5-8) by $\exp(+im\pi x)$, where m is an integer, and integrate over the range $x = 0$ to $x = 2$. After considerable algebraic manipulation, we arrive at

$$f(x) = \frac{4}{\pi}\left(\frac{\sin \pi x}{1} + \frac{\sin 3\pi x}{3} + \frac{\sin 5\pi x}{5} + \frac{\sin 7\pi x}{7} + \cdots\right) \qquad (5\text{-}12)$$

An infinite number of terms are required to represent the function exactly, although a few terms may suffice for a crude approximation. The curves for $f(x) = (4/\pi)\sin \pi x$ and for

$$f(x) = \frac{4}{\pi}\left(\sin \pi x + \frac{\sin 3\pi x}{3} + \frac{\sin 5\pi x}{5}\right)$$

are shown in Fig. 5-10.

EXERCISE 5-8 Use Eq. (5-9) to evaluate: (a) $e^{\pi i n}$, where n is an integer; (b) $e^{3\pi i/2}$; (c) $e^{\pi i/6} - e^{-\pi i/6}$; (d) $\ln(-1)$.

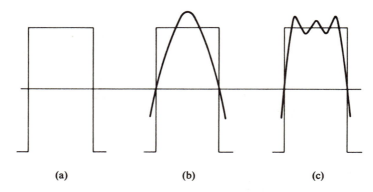

FIG. 5-10 $(a) f(x)$; (b) $4/\pi \sin \pi x$; (c) $4/\pi[\sin \pi x + (\sin 3\pi x)/3 + (\sin 5\pi x)/5]$.

EXERCISE 5-9   Carry out the details of the calculation of the coefficients for the Fourier expansion of the function shown in Fig. 5-9. How could you have predicted that only sine terms would appear in the final result and how could this have been used to simplify the calculation?

5-14   Fourier expansion of electron density

The Fourier expansion of the electron density function is

$$\rho(xyz) = \sum_h \sum_k \sum_l F(hkl) \exp[-2\pi i(hx + ky + lz)] \qquad (5\text{-}13)$$

where $F(hkl)$ is the coefficient to be determined, and h, k, and l are integers over which the series is summed. Because of the three-dimensional periodicity, a triple summation is required here. The coefficients may be evaluated by multiplying both sides by $\exp[2\pi i(h'x + k'y + l'z)]$ and integrating

$$\int_0^1 \int_0^1 \int_0^1 \rho(xyz) \exp[2\pi i(h'x + k'y + l'z)]\, dx\, dy\, dz$$

$$= \int_0^1 \int_0^1 \int_0^1 \exp[2\pi i(h'x + k'y + l'z)] \sum_h \sum_k \sum_l F(hkl)$$

$$\exp[-2\pi i(hx + ky + lz)]\, dx\, dy\, dz \qquad (5\text{-}14)$$

The only nonvanishing term on the right occurs when $h = h'$, $k = k'$, $l = l'$. The result is

$$\int_0^1 \int_0^1 \int_0^1 \rho(xyz) \exp[2\pi i(hx + ky + lz)]\, dx\, dy\, dz = F(hkl) \qquad (5\text{-}15)$$

in which we are ignoring a factor of V, the unit cell volume. The left-hand side of Eq. (5-15) is known as the Fourier transform of the function $\rho(xyz)$. If we knew the value of $\rho(xyz)$ at every point x, y, z, we could evaluate $F(hkl)$ by integrating Eq. (5-15). Knowing $\rho(xyz)$ at every point is tantamount to knowing the crystal structure, so if we knew the crystal structure, we could calculate $F(hkl)$ for all values of h, k, and l. On the other hand, if we knew all of the $F(hkl)$ values, we could calculate the electron density by means of Eq. (5-13).

5-15 Intensities of diffraction spots

The contribution of X-ray diffraction to the solution of the problem is that the intensity of the radiation reflected from the plane (hkl) is proportional to $|F(hkl)|^2$,

$$I(hkl) \propto |F(hkl)|^2 \tag{5-16}$$

There are various other factors that influence the intensity, and derivation of the values of $|F(hkl)|^2$ from measured intensities will require corrections for polarization of X rays, for the length of time the plane is in a reflecting position and, perhaps, for absorption of the X rays by the crystal. The intensities will also be affected by the size of the crystal, by the condition of the crystal, and by thermal vibrations in the crystal structure (see Section 5-18). However, the only dependence on the atomic positions is given by Eq. (5-16) [since $F(hkl)$ can be obtained from the structure by Eq. (5-15)], and a set of relative values of $|F(hkl)|^2$ can routinely be obtained from a set of measured intensities.[7] These intensities may be obtained by measuring the densities of the spots on photographic film or by means of counting methods. The photographic methods may make use of photoelectric densitometers or they may consist of visual estimation of the intensities by comparison with a standard scale. Counter methods may use Geiger, proportional, or scintillation counters.

5-16 The phase problem

Unfortunately, the availability of sets of values of $|F(hkl)|^2$ does not lead to routine determination of crystal structures. Equation (5-13) requires values of $F(hkl)$, whereas the intensities only give $|F(hkl)|^2$.

---

[7] The physical theory of X-ray diffraction is covered in R. W. James, *The Optical Principles of the Diffraction of X-rays*, Bell, London, 1954, and in W. H. Zachariasen, *Theory of X-ray Diffraction in Crystals*, Wiley, New York, 1945 [reprinted by Dover, New York, 1967].

According to Eqs. (5-15) and (5-9), $F(hkl)$ is a complex number. That is, we can write

$$F(hkl) = A(hkl) + iB(hkl) \tag{5-17}$$

where $i = \sqrt{-1}$. However,

$$|F|^2 = (A + iB)(A - iB) = A^2 + B^2$$

(for simplicity we have omitted writing the h, k, and l) so the intensity only gives $A^2 + B^2$, and the values of A and B are not obtained directly. For example, if $|F(hkl)|^2 = 10$, $F(hkl)$ can be any complex number $A + iB$ such that $A^2 + B^2 = 10$, and there are an infinite number of possibilities; a few such numbers are $\sqrt{10}$, $-\sqrt{10}$, $\sqrt{10}i$, $\sqrt{5} + \sqrt{5}i$, $-\sqrt{5} + \sqrt{5}i$, $3 + i$, $3 - i$, $\sqrt{6} + 2i$, and $2 - \sqrt{6}i$. In order to calculate the crystal structure by means of Eq. (5-13) we must know A and B individually. This apparent impasse is known as the *phase problem* in crystallography. Things are not entirely dark, however, since the complete set of intensities provides enough information so that crystal structures can be solved, and methods of solving the phase problem will be treated in Chapter 6.

5-17   Calculation of structure factors

When a crystal structure is known, values of $F(hkl)$ can be calculated, and a test of the correctness of a structure is how well the calculated values of $F(hkl)$ agree with the observed magnitudes. Equation (5-15) is not particularly convenient for calculating $F(hkl)$ values, and we now proceed to derive a more useful form. If we regard atoms as discrete, separated regions of electron density, the function $\rho(xyz)$ will be different from 0 only when the point x,y,z is near an atom, and contributions to the integral of Eq. (5-15) will result only from the regions of space near atoms. We may then write Eq. (5-15) as a sum of integrals

$$F(hkl) = \sum_j \int \int \int \rho(xyz) \exp[2\pi i(hx + ky + lz)] \, dx \, dy \, dz \tag{5-18}$$

where each term in the summation is a triple integral over the volume of a single atom, and the summation is over all the atoms in the unit cell. We now rechoose the origin in each triple integral to be at the center of the particular atom. For atom j, centered at x_j, y_j, z_j, we choose new coordinates

$$x' = x - x_j$$
$$y' = y - y_j$$
$$z' = z - z_j$$

Substitution of these in Eq. (5-18) gives

$$F(hkl) = \sum_j \exp[2\pi i(hx_j + kx_j + lz_j)] \int \int \int \rho(x' y' z')$$
$$\exp[2\pi i(hx' + ky' + lz')] \, dx' \, dy' \, dz' \qquad (5\text{-}19)$$

There is a triple integral in Eq. (5-19) for each atom. It is a quite reasonable approximation that all atoms of a given type will have the same electron distributions, regardless of the compounds in which they occur. There undoubtedly will be differences in the electron arrangement due to the type of bonding, but the X-ray scattering is due to all the electrons in the atom, and variations resulting from slightly different valence states will be minor. This makes it possible to calculate numerical values of these integrals by quantum mechanical methods, and tables of such values are given in Volume II of the *International*

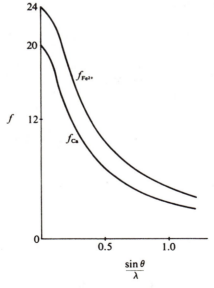

FIG. 5-11   *Plots of atomic scattering factors for Ca and for Fe^{2+}.*

Tables for X-ray Crystallography. These quantities are denoted by the letter f, and they are called *atomic scattering factors* or form factors. The value of f depends on the type of atom and on the Bragg angle θ. Values for the neutral calcium atom and for the Fe^{2+} ion are plotted in Fig. 5-11. A scale is used such that the value of f when $\theta = 0$ is equal to the number of electrons in the atom. When the atomic scattering factor is introduced into Eq. (5-19), we have

$$F(hkl) = \sum_j f_j \exp[2\pi i(hx_j + ky_j + lz_j)] \tag{5-20}$$

Values of $F(hkl)$, the structure factors, may be readily computed by means of this formula.

5-18   Effect of thermal vibration

Calculated structure factors are frequently modified by introducing a temperature factor, which takes into account that the atoms undergo constant vibration about their equilibrium positions. In this case

$$F(hkl) = \sum_j f_j \exp[2\pi i(hx_j + ky_j + lz_j)] \exp\left[-B_j\left(\frac{\sin\theta}{\lambda}\right)^2\right] \tag{5-21}$$

where B_j is proportional to the mean square displacement of atom j from its equilibrium position. Values of B_j can be obtained by comparing the calculated structure factors with the observed magnitudes. In some cases, anisotropic temperature factors are determined, which account for variations with direction of the amplitude of vibration.

5-19   Structure factors of centrosymmetric crystals

Equation (5-20) may be written in the form of Eq. (5-17) by applying Eq. (5-9):

$$\begin{aligned} F(hkl) &= \sum_j f_j \cos 2\pi(hx_j + ky_j + lz_j) \\ &+ i\sum_j f_j \sin 2\pi(hx_j + ky_j + lz_j) \end{aligned} \tag{5-22}$$

If the crystal has a center of symmetry at the origin of the unit cell, then, if there is an atom at x_j, y_j, z_j, there is an equivalent atom at $-x_j, -y_j, -z_j$.

Therefore,

$$F(hkl) = \sum_j f_j [\cos 2\pi(hx_j + ky_j + lz_j) + \cos 2\pi(-hx_j - ky_j - lz_j)]$$
$$+ i \sum_j f_j [\sin 2\pi(hx_j + ky_j + lz_j) +$$
$$\sin 2\pi(-hx_j - ky_j - lz_j)] \qquad (5\text{-}23)$$

where the summation is now over atoms not related by the center of symmetry. Since $\cos(-\phi) = \cos\phi$, and $\sin(-\phi) = -\sin\phi$, we have

$$F(hkl) = 2 \sum_j f_j \cos 2\pi(hx_j + ky_j + lz_j) \qquad (5\text{-}24)$$

We have achieved the valuable result that the structure factor of a centrosymmetric crystal is a real number; that is, the imaginary component involving $i = \sqrt{-1}$ has vanished. This doesn't entirely eliminate the phase problem, since we still must decide whether $F(hkl)$ is positive or negative, but it does vastly simplify things. Unfortunately, nature doesn't always accommodate us by forming centrosymmetric crystals, and this seems to be particularly true as the structures get more complex, as in the case of materials of biological interest.

5-20   *Friedel's law*

We now consider a crystal that does not have a center of symmetry. The structure factor for the plane (hkl) is given by Eq. (5-22), or we may write

$$F(hkl) = A + iB$$

using the abbreviated notation of Eq. (5-17). The structure factor for the plane with indices $-h, -k, -l$ is obtained by changing the signs of the indices in Eq. (5-22). Therefore,

$$F(\bar{h}\bar{k}\bar{l}) = A - iB$$

The observed intensities for the two cases are proportional to

$$|F(hkl)|^2 = A^2 + B^2$$

and

$$|F(\bar{h}\bar{k}\bar{l})|^2 = A^2 + B^2$$

These two planes, therefore, give the same intensities, and the diffraction pattern has a center of symmetry, whether the crystal has one or not.

This is Friedel's law, and as a consequence we cannot usually tell by inspection of a set of photographs whether a crystal has a center of symmetry or not. The structure, once it is derived, will, of course, tell us the true symmetry of the crystal, and there are also statistical methods of detecting a center of symmetry from the distribution of intensities. However, the diffraction patterns will be centrosymmetric.[8]

5-21 Laue groups

The symmetry of a diffraction pattern must be that of one of the centro-symmetric crystallographic point groups. If, for example, we have a crystal whose point group is 4, not only will no difference in the diffraction pattern be detected if we rotate the crystal through 90° about c, but there will also be no difference in intensity between the (hkl) and $(hk\bar{l})$ planes. The diffraction pattern has all the symmetry of point group $4/m$, and directions that are equivalent in $4/m$ will have equal intensities of diffraction.

The diffraction symmetry thus assists us in classifying crystals. If the Laue group is observed to be $4/m$, the crystal system is tetragonal, the point group of the crystal is either 4, $\bar{4}$, or $4/m$, and the space group is one of those associated with these three point groups. Our determination of the crystal system is, therefore, based on the diffraction symmetry, which may be deduced by inspection of a series of photographs. The eleven Laue groups are listed in Table 3-1.

EXERCISE 5-10 (a) Derive the general positions for space group $P4_1$. (b) Calculate the structure factors for the (hkl), $(\bar{k}hl)$, $(\bar{h}\bar{k}l)$, $(k\bar{h}l)$, $(hk\bar{l})$, $(\bar{k}h\bar{l})$, $(\bar{h}\bar{k}\bar{l})$, and $(k\bar{h}\bar{l})$ planes, and show that these eight planes will give equal intensities.

EXERCISE 5-11 The general positions of $P3$ are x,y,z; $\bar{y}, x-y, z$; $y-x, \bar{x}, z$. Show that reflections from the planes (hkl), (ihl), (kil), $(\bar{h}\bar{k}l)$, $(\bar{i}\bar{h}l)$, $(\bar{k}\bar{i}l)$,

---

[8] An exception to Friedel's law occurs in the case of anomalous dispersion. This happens when the X-ray wavelength is such that the X rays are highly absorbed by the atoms in the crystal. Mathematically, the result is that the atomic scattering factors for the atoms concerned are complex numbers. An important application of this effect is in determining absolute configurations of molecules; i.e., distinguishing between enantiomorphs. Anomalous dispersion is also useful in solving the phase problem. See, for example, H. Lipson and W. Cochran, *The Determination of Crystal Structures*, 3rd Ed., Cornell Univ. Press, Ithaca, New York, 1966, Chapter 14. (Also published by Bell, London, 1966.)

where $i = -h - k$, have equal intensities. Note: Four indices, $hkil$, are sometimes used in the trigonal and hexagonal systems, so that the indices of equivalent reflections can quickly be generated by permuting h, k, and i.

5-22 Structure factors of sodium chloride

The sodium chloride structure (Fig. 1-1) is face-centered cubic with

Na^+ ions at $0, 0, 0$; $\frac{1}{2}, \frac{1}{2}, 0$; $\frac{1}{2}, 0, \frac{1}{2}$; and $0, \frac{1}{2}, \frac{1}{2}$; and

Cl^- ions at $\frac{1}{2}, 0, 0$; $0, \frac{1}{2}, 0$; $0, 0, \frac{1}{2}$; and $\frac{1}{2}, \frac{1}{2}, \frac{1}{2}$.

$$F(hkl) = f Na^+ \left(\exp[2\pi i(0)] + \exp\left[\frac{2\pi i(h+k)}{2}\right] + \exp\left[\frac{2\pi i(h+l)}{2}\right] \right.$$
$$\left. + \exp\left[\frac{2\pi i(k+l)}{2}\right] \right) + f Cl^- \left(\exp\left[\frac{2\pi ih}{2}\right] + \exp\left[\frac{2\pi ik}{2}\right] \right.$$
$$\left. + \exp\left[\frac{2\pi il}{2}\right] + \exp\left[\frac{2\pi i(h+k+l)}{2}\right] \right)$$

Now, $\exp(2\pi in/2) = \cos \pi n + i \sin \pi n$ and, if n is an integer, this reduces to $\cos \pi n = (-1)^n$. That is, $\exp(2\pi in/2)$ is $+1$ if n is an even integer and -1 if n is an odd integer. The structure factor formula can be reduced to

$$F(hkl) = [1 + (-1)^{h+k} + (-1)^{k+l} + (-1)^{h+l}][f Na^+ + (-1)^h f Cl^-]$$

If we examine some possible values of the indices, we arrive at Table 5-1.

TABLE 5-1   STRUCTURE FACTORS OF NaCl

hkl	$F(hkl)$
100	0
110	0
111	$4(f Na^+ - f Cl^-)$
200	$4(f Na^+ + f Cl^-)$
210	0
211	0
220	$4(f Na^+ + f Cl^-)$
300	0
221	0
310	0
311	$4(f Na^+ - f Cl^-)$
222	$4(f Na^+ + f Cl^-)$

The term $[1 + (-1)^{h+k} + (-1)^{h+l} + (-1)^{k+l}]$ is 0 unless the indices are either all odd or all even, and the structure factor is, therefore, zero for planes with mixed indices such as (110) or (210). This is characteristic of face centering, and the presence of face centering can, therefore, be detected by the systematic absence of reflections of a certain type from the diffraction pattern. Similar considerations apply to other types of centering, and Table 5.2 summarizes the conditions under which reflections appear.

EXERCISE 5-12 Derive the condition on the allowed Miller indices for a body-centered crystal.

EXERCISE 5-13 Prepare a table such as Table 5-1, for the CsCl structure, where the unit cell contains one Cs^+ at $0,0,0$ and one Cl^- at $\frac{1}{2},\frac{1}{2},\frac{1}{2}$. What is the lattice type in this case?

EXERCISE 5-14 Potassium chloride has the same structure as NaCl. Rewrite Table 5-1 for the case of KCl, assuming that, since K^+ and Cl^- are isoelectronic, they have the same atomic scattering factors.

5-23 Extinctions due to glide planes

Suppose we have a c glide plane perpendicular to the b axis. For an atom at x,y,z, there is an equivalent atom at $x,\bar{y},\frac{1}{2}+z$. The contribution of these two atoms to the structure factor is

$$F(hkl) = \{\exp[2\pi i(hx + ky + lz)] + \exp[2\pi i(hx - ky + l/2 + lz)]\}f$$

TABLE 5-2   CONDITIONS ON INDICES FOR
APPEARANCE OF GENERAL REFLECTIONS

Lattice type	Condition
P	None
I	$h + k + l = 2n$, where n is an arbitrary integer
F	$h + k = 2n$, $k + l = 2n$, $h + l = 2n$ (indices all even or all odd)
A	$k + l = 2n$
B	$h + l = 2n$
C	$h + k = 2n$
R	$-h + k + l = 3n$

For the special case of $k = 0$,

$$F(h0l) = \exp\left[2\pi i(hx + lz)\right]\left[1 + \exp\left(\frac{2\pi il}{2}\right)\right]f$$

$$= \exp\left[2\pi i(hx + lz)\right][1 + (-1)^l]f$$

$$= 0 \text{ if } l \text{ is odd,}$$

$$= 2\exp\left[2\pi i(hx + lz)\right] \text{ if } l \text{ is even}$$

Reflections of the type $(h0l)$ will, therefore, be missing unless l is an even number. The characteristic absence or extinction of $(h0l)$ reflections with l odd thus indicates a c glide plane perpendicular to b, and such extinctions are extremely useful in deducing the space group of an unknown crystal.

5-24 *Extinctions due to screw axes*

We consider for an example a twofold screw axis parallel to b. The equivalent positions related by the screw axis are x, y, z and $\bar{x}, \frac{1}{2} + y, \bar{z}$. Thus,

$$F(0k0) = e^{2\pi iky}[1 + (-1)^k]f$$

and $(0k0)$ reflections will be absent unless k is an even integer.

The extinctions due to the various types of glide planes and screw axes are given in Volume I of the *International Tables for X-ray Crystallography*, along with tables to aid in deducing space groups.

EXERCISE 5-15 Deduce all conditions for general and special reflections for the following space groups: (a) $C2/c$; (b) *Aba2*; (c) *Imma*.

EXERCISE 5-16 The structure of coesite, a high-pressure form of SiO_2, is monoclinic. The unit cell has $a = 7.17$ Å, $b = 12.38$ Å, $c = 7.17$ Å, $\beta = 120°$, and contains sixteen SiO_2 groups. A complete structure determination has verified the monoclinic symmetry and established the space group as $C2/c$.

(a) Discuss the apparent hexagonal unit cell dimensions. Is it possible for a C-centered cell with these dimensions to be hexagonal?

(b) A body-centered cell may be obtained by choosing vectors from the origin to the points $\frac{1}{2}, \frac{1}{2}, 1$; $\frac{1}{2}, -\frac{1}{2}, 1$; and $1, 0, 0$. Calculate the lengths of the three edges and the three angles of this new unit cell, and show that it is dimensionally nearly tetragonal.

(c) Suggest an X-ray diffraction photograph that will show whether or not the crystal system is actually tetragonal.

EXERCISE 5-17 Precession and Weissenberg photographs of a crystal indicated a unit cell of dimensions $a = 15.97$ Å, $b = 15.97$ Å, $c = 42.47$ Å, $\alpha = \beta = \gamma = 90°$. There are lattice points at $0,0,0$; $\frac{3}{4},\frac{1}{4},\frac{1}{4}$; $\frac{1}{2},\frac{1}{2},\frac{1}{2}$; and $\frac{1}{4},\frac{3}{4},\frac{3}{4}$.

(a) Determine the conditions for general reflections. That is, what classes of reflections are systematically missing?

(b) Do the unit cell dimensions indicate the tetragonal crystal system?

(c) The space group of this crystal is actually $C2/m$. Choose a C-centered monoclinic cell in which the unique axis b is given by the vector from the origin to the point $1,1,0$.

Chapter 6

DETERMINATION OF
ATOMIC POSITIONS

In Chapter 5 we saw how X-ray diffraction provides a means for determining the size and shape of the unit cell. The Laue group can be obtained from the symmetry of the diffraction patterns, and the lattice type can be deduced from systematic absences or extinctions among general reflections. Extinctions of special types of reflections indicate the presence of glide planes and screw axes, and such observations aid in deducing the space group, although a unique choice of space group cannot usually be made on the basis of these data alone. For example, space groups *P*222, *Pmm*2, and *Pmmm* could not be distinguished without more information.

In Chapter 5 we also learned how the intensities of the Bragg reflections are related to the atomic positions, and we encountered the phase problem which prevents us from proceeding automatically from the measured intensities to the structure. In this chapter we will briefly discuss some of the methods that make crystal structure determination possible. The phase problem is not an insurmountable obstacle because the quantity of data available usually vastly exceeds the number of parameters to be determined. Each atom to be located in the unit cell requires the specification of three coordinates, and one temperature factor is usually also required. (If anisotropy of the thermal vibrations is important, six components of the temperature factor may be required.) We, thus, need to determine a minimum of four parameters per atom, and a structure involving twenty atoms would involve at least eighty parameters. However, we would probably measure at least 10 times this number of reflections, so that the problem is greatly overdetermined.

6-1   Solutions of structure factor equations

It might seem possible, in principle, to write Eq. (5-21) for each measured reflection and to solve the set of simultaneous equations for the unknown parameters. With one observation for each parameter, the problem would seem to have a solution. Of course, our intensity measurements are subject to error, so perhaps it would be preferable to include all of our observations and seek a solution by means of some least squares process. One difficulty is that we know only $|F(hkl)|$ rather than $F(hkl)$. A possible response to this complication is that we will square Eq. (5-21), so as to obtain a set of equations for the positive quantities $|F(hkl)|^2$. However, these equations are hopelessly intractable, and no one has yet succeeded in solving such sets of simultaneous equations with a large number of unknowns.

It would be well at this stage also to dispose of the possibility of obtaining a trial-and-error solution by evaluating the Fourier series for all possible combinations of signs. Only a simple structure could be solved with 100 terms in the Fourier series, and there would be 2^{100} different sign combinations. This number is approximately 10^{30}, and you might like to calculate the number of centuries that correspond to 10^{30} seconds.

The methods by which the required information can be extracted from the intensity data are much more subtle than these brute force attempts, and the search for new and improved methods is still a very active field of research.

6-2   The Patterson function

In 1934, A. L. Patterson discovered that a Fourier series using values of $|F(hkl)|^2$ as coefficients instead of $F(hkl)$ could produce useful information about the structure. To derive Patterson's function, we take the electron density at point x,y,z, given by Eq. (5-13), and multiply it by the electron density at the point $x + u, y + v, z + w$. That is, we form the product

$$\rho(xyz)\,\rho(x + u, y + v, z + w)$$

We now multiply by $dx\,dy\,dz$ and integrate over the volume of the unit cell,

$$\int_0^1 \int_0^1 \int_0^1 \rho(xyz)\,\rho(x + u, y + v, z + w)\,dx\,dy\,dz$$

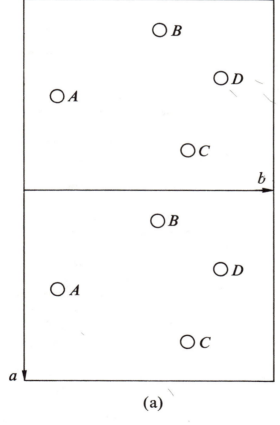

(a)

FIG. 6-1   (a) *Two unit cells of a structure containing four atoms* (*continued on p. 116*).

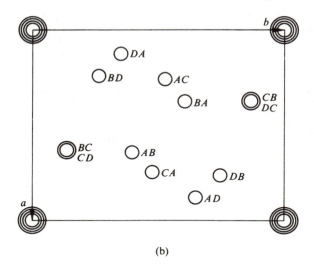

(b)

FIG. 6-1   (*b*) *Vector map of the structure of* (*a*).

When we substitute Eq. (5-13) for each electron density function, we eventually arrive at

$$P(uvw) = \sum \sum \sum |F(hkl)|^2 \cos 2\pi(hu + kv + lw) \qquad (6-1)$$

The Patterson function $P(uvw)$ will be nonzero at the point u, v, w only when there exist points x, y, z such that $\rho(xyz)$ and $\rho(x + u, y + v, z + w)$ are both nonzero. The Patterson function will reach maximum values at points u, v, w, which correspond to the coordinates of vectors between pairs of atoms. For example, if a crystal structure has an atom at $0.20, 0.31, 0.33$ and another atom at $0.15, 0.18, 0.22$, there will be a maximum in the Patterson function at the point $0.20-0.15, 0.31-0.18$, $0.33-0.22 = 0.05, 0.13, 0.11$. The Patterson function thus gives a map of the vectors between atoms, and there is a Patterson peak for each interatomic vector. As an example, the four-atom structure of Fig. 6-1a gives the vector map of Fig. 6-1b. As a result of the periodicity of the structure, one unit cell contains one vector of each type, and the vector from an A atom to a C atom is related by a lattice translation to every other AC vector. For each vector, such as AC, there is a corresponding vector in the reverse direction, CA, so the Patterson map has a center of symmetry, even though the actual structure in this case is noncentrosymmetric. In this example, the vectors BC and CD happen to coincide. For the particular point $0, 0, 0$, Eq. 6-1 gives

$$P(000) = \sum \sum \sum |F(hkl)^2| \tag{6-2}$$

so the Patterson function has a large positive value at the origin, corresponding to the vectors from each atom to itself.

EXERCISE 6-1 Construct a vector map, similar to Fig. 6-1b, for the following two-dimensional set of points:

x	y
0.11	0.13
0.15	−0.05
0.39	0.56
−0.24	0.22
0.60	0.35

It is possible, in principle, to deduce the atomic arrangement from the vector diagram. Figure 6-1a can thus be derived from Fig. 6-1b, except for the choice of origin, which is arbitrary, and except for inversion of the structure through the center, which would give an enantiomorphic structure with the same values of $|F(hkl)|^2$. The Patterson function is, therefore, an extremely powerful aid to structure determination, and a crystal structure investigation will usually include calculating this function at a large number of points throughout the unit cell so that the coordinates of points where $P(uvw)$ is large can be found. Nevertheless, the interpretation of the Patterson function is not always straightforward, and by itself it does not provide a general solution of the phase problem. One difficulty with the Patterson function is that there are so many interatomic vectors. If a unit cell contains N atoms, there are N^2 vectors. Of these, N are the origin vectors, so there are $N^2 - N$ peaks other than the origin peak. Half of these are related to the other half by a center of symmetry (peak CA in Fig. 6-1b is related to peak AC by inversion through the center of symmetry), so there are $(N^2 - N)/2$ independent peaks. If N is 20, there are 190 independent Patterson peaks, and the vector map will be very crowded. A further complication is that the atoms are not points, but they occupy a considerable volume. In fact, the N atoms pretty well fill up a unit cell. There will also be a range of values at which $P(uvw)$ is different from zero. This is illustrated in one dimension in Fig. 6-2, where the vector function corresponding to Gaussian atoms is $\sqrt{2}$ times as wide as the atomic peaks. Since the N atoms themselves filled

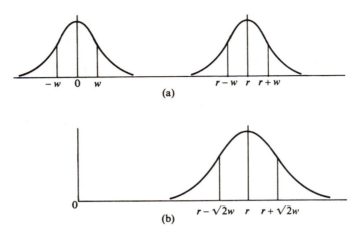

FIG. 6-2   (*a*) *Electron density distribution of atoms of Gaussian shape and width 2w, separated by distance r.* (*b*) *Vector function for atoms of* (*a*) *has width* $2\sqrt{2}w$.

up the unit cell, it is apparent that the N^2 Patterson peaks, each of which is wider than an atom, will overlap considerably. As the structure gets more complex, the situation rapidly gets worse. With N atoms in volume V, there are N^2/V Patterson peaks per unit volume; with $2N$ atoms in volume $2V$, there are $(2N)^2/2V = 2N^2/V$ Patterson peaks per unit volume. The density of Patterson peaks, thus, increases as the complexity of the structure increases, and there is correspondingly less chance of resolving the peaks. Although we could deduce the atomic arrangement if we knew the vector arrangement, the sad fact is that very often we are not able to recognize the individual Patterson vectors.

EXERCISE 6-2   A vector map of a two-dimensional three-atom structure contained, besides the origin peak, independent peaks at $0.85, 0.52$; $0.55, 0.21$; $0.30, 0.31$. Draw a diagram of the vector map, and deduce an atomic arrangement that will account for it.

6-3   Heavy-atom methods

If a unit cell contained only a few atoms, we could deduce the structure from the locations of peaks on the Patterson map. Interpretation of the Patterson function may also be possible for complicated structures if a few of the atoms have appreciably higher atomic numbers than the others. Since the atomic scattering factors increase with the number of electrons on the atoms, the heavy atoms will contribute more to the structure factors, and the Patterson peaks due to the heavy atoms alone may be discernible. In this case, the positions of the heavy atoms may be obtained, and their contributions to the structure factors may be calculated. If the heavy atoms are heavy enough, they may by themselves determine enough phases so that a Fourier map of the electron density will reveal the positions of some of the lighter atoms. If the structure has a center of symmetry, all that is necessary is that the heavy-atom contribution give the correct sign of a sufficient number of structure factors. As an example, a total of 1828 independent reflections were observed from a crystal of $B_{10}H_{12}[S(CH_3)_2]_2$. The sulfur atom positions were readily deduced from the Patterson maps, and the contributions of the sulfur atoms to the 659 largest structure factors were calculated. Of these, the 248 structure factors with the largest sulfur contributions were selected and used in computing an electron density map with the observed magnitudes and calculated signs. This Fourier map indicated positions for two of the carbon atoms and six of the boron atoms of a molecule. Another structure factor calculation included the contributions of these atoms in addition to the sulfur atoms, and a second Fourier calculation based on 448 terms revealed the rest of the carbon and boron atoms. Two of the original 248 signs were eventually shown to have been incorrect.

The success of the heavy-atom method hinges on the presence of an atom heavy enough to determine correctly the phases of a substantial number of structure factors with large magnitudes. Enough phases are required so that the incomplete Fourier series will show some additional atoms. This Fourier calculation will also produce spurious peaks, and a knowledge of structural chemistry is required to distinguish the atoms from the effects of an incomplete series. On the other hand, it is desirable that the heavy atom not to be too heavy or it may dominate the scattering so much that the lighter atoms cannot be recognized on the Fourier maps.

6-4 Isomorphous replacement

Two substances with the same crystal structure are said to be *isomorphous*. For example, the triethyl ammonium halides, $(C_2H_5)_3NHCl$, $(C_2H_5)_3NHBr$, and $(C_2H_5)_3NHI$ have the same atomic arrangement, with only slight differences in unit cell dimensions. The structure factors for these three structures differ mainly in the contributions of the halogen atoms. In this particular example, the structure could be determined by straightforward application of the heavy-atom method. However, the heavy-atom method might not suffice for more complicated structures, but information about the phases could be obtained by comparing the intensities from crystals of the different compounds. If the intensity of a given reflection in a centrosymmetric structure increases in going from the chlorine to the bromine compound, the structure factor must have the same sign as the halogen contribution. If the structure is not centrosymmetric, we need to determine phases rather than merely signs of structure factors, but this information can be deduced from the observed intensities from two or more compounds.

The method of isomorphous replacement is exceedingly important in solving complex structures. The basic requirement is that it be possible to prepare isomorphous derivatives of a compound in which

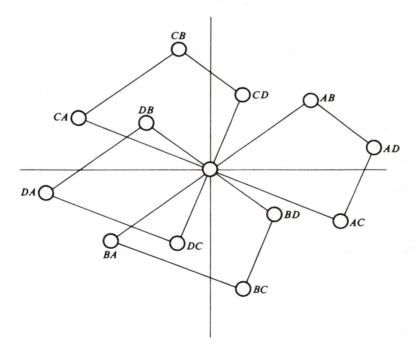

FIG. 6-3   *Arrangement of N atoms repeated N times generates the Patterson map. (See Fig. 6-1.)*

one atom has different scattering power. Besides replacing one halogen atom by another, or replacing hydrogen by a halogen, sulfur may sometimes be replaced by selenium, or a heavy metal atom may be incorporated in the molecule. Among the notable successes of this method are the solutions of the structures of the proteins, hemoglobin and myoglobin, by Kendrew and Perutz. As many as seven isomorphous compounds were used in this work.[1]

6-5   Superposition methods

A graphical method of constructing a Patterson map is indicated in Fig. 6-3. One unit cell of the arrangement of atoms of Fig. 6-1a is used, and the Patterson map is derived by drawing this four-atom structure 4 times, once with each atom at the origin. Figures 6-1b and 6-3 are equivalent if the periodicity of the diagrams is taken into account; for instance, if the point DA in Fig. 6-3 is repeated one unit cell to the right, it coincides with DA of Fig. 6-1b. The important aspect of the Patterson map here is that it contains the structure, repeated N times, where N is the number of atoms per unit cell. A method of recovering the atomic arrangement from the vector map is shown in Fig. 6-4. The diagram in Fig. 6-4 is just four unit cells of the Patterson map of the structure of Fig. 6-1a. This Patterson map has been drawn twice in Fig. 6-4, but the origin of one of the maps has been shifted so that it coincides with one of the nonorigin peaks of the other. Positions where peaks in the two maps coincide outline the structure. Actually, both the structure and its inverse image or enantiomorph are obtained by this process, as a consequence of Friedel's law. This method of superposition is generally applicable to the recovery of a set of points from its vector set.

When the Patterson peaks are not resolved from each other, as is usually the case in crystal structure analysis, the recovery of the atomic arrangement from the Patterson map is not automatic. A procedure suggested by M. J. Buerger involves calculating the Patterson function at a large number of points, superimposing two maps with the origin of one coinciding with a suitable peak of the other, and preparing a new map where the value at each point is the lower of the two superimposed values. This determination of a *minimum function* is essentially the procedure we carried out in Fig. 6-4, where we retained only those points

---

[1] A more complete discussion of heavy-atom and isomorphous replacement methods, including numerous examples and literature references, may be found in M. J. Buerger, *Crystal-Structure Analysis*, Wiley, New York, 1960.

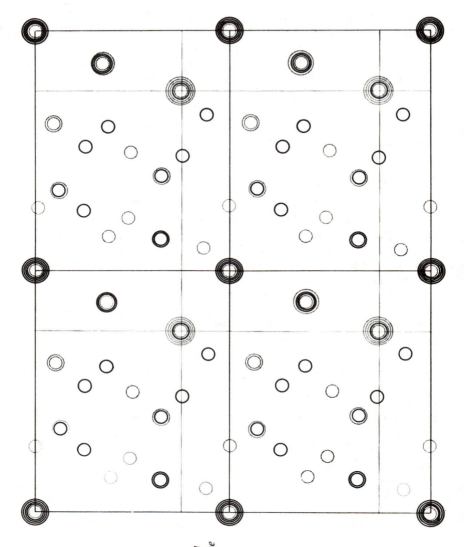

FIG. 6-4 *Superposition of Patterson maps of the structure of Fig. 6-1a. Coincidences include structure.*

that corresponded to peaks on both Patterson maps. The minimum function has far fewer peaks than the Patterson function, since only peaks appearing simultaneously on both maps are retained, and the minimum function may closely resemble the actual structure. This method has been extremely successful on some rather complicated structures. Direct Patterson interpretation is very difficult with ten or more independent atoms of similar atomic number to be determined. As one of many examples in the literature, a modification of the minimum function method led to the solution of the structure of cellobiose, $C_{12}O_{11}H_{22}$, which has twenty-three independent atoms other than hydrogen.[2]

6-6 *Inequalities*

We now consider a few of the so-called *direct methods*, which attempt to determine the phases of the structure factors without first deriving a set of atomic positions.

We have already pointed out the redundancy of a structure determination; that is, there are many more observations than there are parameters, so that the structure factors cannot all be independent. In 1948, D. Harker and J. S. Kasper derived some inequality relationships between the structure factors. We first define a unitary structure factor as

$$U(hkl) = \frac{F(hkl)}{\sum f_j} \tag{6-3}$$

which reduces to

$$U(hkl) = \sum n_j \cos 2\pi(hx_j + ky_j + lz_j) \tag{6-4}$$

where

$$n_j = \frac{f_j}{\sum f_j} \tag{6-5}$$

if the structure has a center of symmetry. A mathematical inequality due to Cauchy is

$$|\sum a_j b_j|^2 \leqslant (\sum |a_j|^2)(\sum |b_j|^2) \tag{6-6}$$

---

[2] An extensive treatment of superposition methods, including the minimum function, may be found in M. J. Buerger, *Vector Space*, Wiley, New York, 1959.

If, for example, $a_1 = 1$, $a_2 = 2$, $b_1 = 3$, $b_2 = 4$, we have

$$(1 \times 3 + 2 \times 4)^2 \leqslant (1 + 4)(9 + 16)$$

or

$$121 \leqslant 125$$

We apply the Cauchy inequality to the unitary structure factor by letting $a_j = (n_j)^{1/2}$, and $b_j = (n_j)^{1/2} \cos 2\pi(hx_j + ky_j + lz_j)$.

$$\left| \sum n_j \cos 2\pi(hx_j + ky_j + lz_j) \right|^2$$
$$\leqslant \sum n_j \sum n_j \cos^2 2\pi(hx_j + ky_j + lz_j) \tag{6-7}$$

From Eq. (6-5), $\sum n_j = 1$, and from the trigonometric identity

$$\cos^2 A = \frac{1 + \cos 2A}{2},$$

we have

$$|U(hkl)|^2 \leqslant \tfrac{1}{2} \sum n_j[1 + \cos\{2\pi \times 2(hx_j + ky_j + lz_j)\}] \tag{6-8}$$

$$|U(hkl)|^2 \leqslant \tfrac{1}{2}[1 + \sum n_j \cos 2\pi(2hx_j + 2ky_j + 2lz_j)] \tag{6-9}$$

$$|U(hkl)|^2 \leqslant \tfrac{1}{2}[1 + U(2h, 2k, 2l)] \tag{6-10}$$

To illustrate the use of Eq. (6-10), suppose $|U(130)| = 0.50$ and $|U(260)| = 0.60$. The sign of $U(260)$ must be $+$ in order to satisfy the inequality, and we, therefore, have deduced the phase of the (260) reflection. Inequality relationships appropriate to a given space group can be derived by symmetry considerations. It has been shown that these inequalities are all a consequence of the simple fact that the electron density function $\rho(xyz)$ is never less than zero. That is, the only mathematical requirement is that the number of electrons per unit volume cannot be negative, although the nature of the inequalities, and their utility in phase determination, depends upon symmetry.

Another inequality for centrosymmetric structures is

$$U^2(hkl) + U^2(h'k'l') + U^2(h+h', k+k', l+l')$$
$$\leqslant 1 + 2U(hkl)\, U(h'k'l')\, U(h+h', k+k', l+l') \tag{6-11}$$

An illustration of Eq. (6-11) is

$$U^2(123) + U^2(021) + U^2(144) \leqslant 1 + 2U(123)\, U(021)\, U(144)$$

If the unitary structure factors are all large, this implies that $U(144)$ has the same sign as the product $U(123)\, U(021)$. Such inequalities are useful for determining more signs such as that of $U(144)$, once some signs, such as those of $U(123)$ and $U(021)$, are known.

The Harker–Kasper inequalities were first used in determining the structure of decaborane, $B_{10}H_{14}$. Unfortunately, they apply only when the magnitudes of the unitary structure factors are large. The maximum value of $U(hkl)$ is 1, and statistical analysis shows that the average value of $|U(hkl)|^2$ is $1/N$, where N is the number of atoms in the unit cell. As the complexity of the structure increases, therefore, the fraction of the structure factors for which the inequality relationships are applicable declines.

6-7 Sayre–Cochran–Zachariasen relationship

The result that the inequality in Eq. (6-11) enables us to deduce the sign of $U(h + h', k + k', l + l')$ from the known signs of $U(hkl)$, and $U(h'k'l')$ may be expressed symbolically as

$$S(h + h', k + k', l + l') = S(hkl)\, S(h'k'l') \qquad (6\text{-}12)$$

which says that the sign of the structure factor with indices $h + h', k + k', l + l'$ is the product of the signs of $F(hkl)$ and $F(h'k'l')$. If the structure factors are not large, we cannot guarantee the truth of Eq. (6-12). However, even if the magnitudes of the structure factors are not quite sufficient to apply the inequality, Eq. (6-12) will probably be true. Slight shifts of some of the atoms might be enough to increase the three structure factors so that Eq. (6-11) could be used, and it is unlikely that these shifts would actually change the signs of the structure factors. The probability that Eq. (6-12) will give the correct sign for $F(h + h', k + k', l + l')$ depends upon the magnitudes of the structure factors involved and varies from 1 (complete certainty) when (6-11) is satisfied to $\frac{1}{2}$ (complete uncertainty) when one or more of the values is zero.

Once a few signs have been determined, Eq. (6-12) can be used to generate more signs, and these, in turn, can be combined to produce more. So long as structure factors with large magnitudes are considered, the generated signs will probably be correct, and these large structure factors are the ones required to produce a recognizable Fourier representation of the structure.

This is the essence of the method developed independently by D. M. Sayre, W. Cochran, and W. H. Zachariasen, with contributions from many others, and it is frequently referred to as the SCZ method. The

SCZ method has been applied successfully to numerous structures, and computer programs have been prepared that will develop a set of signs consistent with Eq. (6-12).

6-8   *Hauptman–Karle methods*

We have seen that Eq. (6-12) has a probable validity. The next problem is to evaluate numerically the probability that a certain structure factor is positive. Formulas have been derived by H. Hauptman and J. Karle, among others, for the joint probability distribution of structure factors, from which the probability that a given sign is plus can be calculated, and numerous structures have been solved by these methods. An early impressive illustration of the power of these methods was the determination of the structure of p,p'-dimethoxybenzophenone, $C_{15}O_3H_{28}$, which contained two crystallographically independent molecules, necessitating locating thirty-six atoms other than hydrogen.

6-9   *Summary of phase-determining methods*

We have discussed only a few methods of phase determination; our selection was based somewhat on the methods most commonly encountered in the literature. Other methods of solving the phase problem may be found in more advanced textbooks.[3]

We wish to emphasize that there is no one "best" method. Each crystal structure is different, and each presents its own peculiar problems. Very simple structures may be solved by trial and error. If only a few atoms are involved, the Patterson function may be interpreted directly. The presence of heavy atoms may make more sophisticated procedures unnecessary. Very complicated structures may require the chemical labor of preparing isomorphous derivatives. Both the minimum function and the statistical methods have achieved remarkable results, and they have greatly expanded the range of structures that can be solved. The question as to whether the minimum function or the statistical procedure is better cannot be answered. They represent merely different ways of extracting the information inherent in the measured intensities, and the crystallographer confronted with the phase problem will wish them both continued prosperity and development.

[3] See, for example, H. Lipson and W. Cochran, *The Determination of Crystal Structures*, 3rd Ed., Cornell Univ. Press, Ithaca, New York, 1966. (Also published by Bell, London, 1966.)

6-10   *Refinement*

After the phase problem has been overcome, the crystallographer has available a set of parameters specifying the location of every atom within the unit cell. Slight variations in these parameters will produce variations in the values of the structure factors calculated by means of Eq. (5-21). The best set of parameters is that which will produce the most accurate values of interatomic distances and bond angles. It is hoped that these best parameters are also those that give the best agreement between the calculated and observed structure factor magnitudes. A convenient measure of the correctness of a structure is given by the residual or R value

$$R = \frac{\sum |F_o - F_c|}{\sum |F_o|} \qquad (6\text{-}13)$$

An R of 0.20 may indicate a correct structure, with the best possible values of the atomic parameters used in calculating the F_c values. Usually, R should be considerably less than 0.20, unless the reason for the poor agreement is understood. On the other hand, there may be something wrong, such as a light atom incorrectly placed, with a structure that gives an R of 0.10 if the quality of the crystal and the data were high. Generally, an R of 0.30 indicates that the correct structure is near, and if R is 0.10 or less the results are probably very reliable. Among the advantages of structural determination by means of X-ray diffraction is the great confidence that may usually be placed in the structure.

The convincing arguments for the correctness of a crystal structure are the chemical reasonableness of the structure (no inexplicable bond lengths or angles) and the agreement between the values of F_c and F_o. Journals such as *Acta Crystallographica, Zeitschrift für Kristallographie*, and *Inorganic Chemistry* usually publish tables of these calculated and observed structure factors. A correct structure should have generally good agreement, indicated by a low R, with no major unexplained discrepancies, and no special classes of reflections that are bad.

There are numerous sources of error in the measured intensities, and it may not be feasible to correct the data for all of these. Some of the common errors are due to absorption of the X rays by the crystal, diminution of the beam by scattering, and background radiation.

The calculated structure factors may suffer from inaccurate or inappropriate scattering factors, neglect of anisotropic thermal motion, and inability to locate hydrogen atoms.

Once all the phases have been determined, a Fourier map may be calculated, and the atomic positions may be taken as the locations of the maxima of the electron density function. The advent of high-speed computers has led to widespread use of the method of least squares which automatically adjusts the parameters so as to minimize some function such as $\sum (F_o - F_c)^2$. The least squares method gives estimates of the standard deviations of the atomic parameters, which can be used to calculate the uncertainties in bond lengths and angles.

Chapter 7
SOME SIMPLE
STRUCTURES

The principles we have developed throughout this book can be applied to structures of great complexity. In this chapter we want to describe the structures of some of the elements and of a few simple compounds. Our purpose will not be so much to illustrate the material of the preceding chapters as it will be to gain an understanding of some elementary structural concepts. The structures we will consider will have high symmetries, and frequently we will regard these structures as assemblages of closely packed spheres. Crystals composed of molecules cannot be expected to have these high symmetries, since the molecules themselves are of low symmetry. Nevertheless, a detailed acquaintance with some simple structures will help us in interpreting the arrangements in more complicated substances.

7-1 Close packing

The structures of several elements may be described as close-packed arrangements of identical spheres. The only possible close-packed arrangement of spheres in two dimensions is shown in Fig. 7-1; each

FIG. 7-1   *Close packing of spheres in two dimensions.*

sphere is tangent to six other spheres. Three-dimensional close packing is achieved by stacking layers of this type on top of each other, so that each sphere nestles into the interstices of the layers above and below it. Each sphere then has twelve neighbors, six within its own layer, three in the layer above, and three in the layer below. No one has succeeded in proving that this is actually the most compact arrangement of spheres in three dimensions, but it is the closest packing possible if the resulting arrangement is to be periodic.

FIG. 7-2   *Two close-packed layers, with upper layer stacked above interstices of lower layer.*

7-2 Cubic close packing

Figure 7-2 shows portions of two close-packed layers stacked together. The spheres of the upper layer are above some of the interstices of the layer below. If a third layer is stacked on top of the second layer of Fig. 7-2, there are two places it can go. It will of course, have to fit into the interstices of the second layer, but its spheres can be either directly above the spheres of the first layer or directly above a set of interstices of the first layer. The latter alternative is shown in Fig. 7-3.

EXERCISE 7-1 Obtain fourteen identical spheres (Styrofoam balls or Ping-pong balls are satisfactory). Make two close-packed triangles, using six balls in each and stack the triangles together as shown in Fig. 7-2. Use glue to hold the balls together. Add a thirteenth sphere, as shown in Fig. 7-3, so that it is above interstices of the two layers below. Add the fourteenth sphere directly below the thirteenth, so that your model has portions of four layers. Verify that the resulting model is a face-centered cube.

This sequence of close-packed layers gives a periodic structure; the structure has cubic symmetry, the cubic unit cell is face centered, and

FIG. 7-3 *Stacking sequence for cubic close-packed structure, showing cubic unit cell.*

the space group is *Fm3m*. The unit cell is shown in Fig. 7-3, where the view is down one of the threefold axes of the cubic cell. We prepared this model by stacking layers in one direction; we imposed one threefold rotation axis in constructing the model. The great regularity of the arrangement led to three other directions equivalent to this threefold axis.

Table 7-1 lists some elements that have cubic close-packed structures. It should be observed that these are all elements that do not readily form polyatomic molecules; they are either metals or inert gases. The atoms of these elements pack together like rigid spheres.

EXERCISE 7-2   Calculate the density of solid krypton at 58°K from the cell dimension in Table 7-1.

TABLE 7-1   CELL DIMENSION OF
CUBIC CLOSE-PACKED ELEMENTS

Element	(Å)
Ac	5.311
Ag	4.086
Al	4.050
Am	4.894
Ar	5.256 (4.2°K)
Au	4.078
Ca	5.575
Ce	5.161
Cu	3.615
Ir	3.839
Kr	5.721 (58°K)
Ne	4.429 (4.2°K)
Ni	3.524
Pb	4.950
Pd	3.890
Pt	3.923
Rh	3.803
Sr	6.086
Th	5.084
Xe	6.197 (58°K)
Yb	5.486

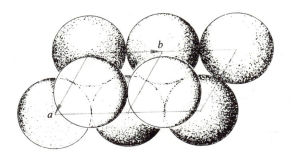

FIG. 7-4   *Two unit cells of hexagonal close-packed structure.*

Since this structure represents the close packing of spheres, it is interesting to compute how efficient the packing is. In other words, how much space is wasted in the interstices between the spheres? If the radius of a sphere is r, the unit cell dimension is $2\sqrt{2}r$ (verify this). The unit cell volume is, therefore, $(2\sqrt{2}r)^3 = 16\sqrt{2}r^3$. The volume of a spherical atom is $\frac{4}{3}\pi r^3$, and the volume actually occupied by the four atoms of the unit cell is $4(\frac{4}{3}\pi r^3) = 16\pi r^3/3$. The fraction of the total space occupied by the atoms is

$$\frac{V_{\text{atoms}}}{V_{\text{cell}}} = \frac{16\pi r^3/3}{16\sqrt{2}r^3} = \frac{\pi}{3\sqrt{2}} = 0.7405$$

In this most compact arrangement of spheres, 25.95% of the space is vacant.

7-3   *Hexagonal close-packed structure*

The other common sequence of close-packed layers is just a repetition of Fig. 7-2. The third layer is directly above the first, the fourth above the second, and so on. Two unit cells of this structure are shown in Fig. 7-4. The unit cell is hexagonal, and the space group is $P6_3/mmc$. The unit cell dimensions in terms of the radius of a sphere are $a = b = 2r$, $c = 4\sqrt{2}r/\sqrt{3}$, $c/a = 2\sqrt{2}/\sqrt{3} = 1.633$.

EXERCISE 7-3   Explain the meaning of each component in the space group symbol $P6_3/mmc$.

EXERCISE 7-4   Prove that $c/a = 2\sqrt{2}/\sqrt{3}$ for the hexagonal close-packed structure.

EXERCISE 7-5   Calculate the efficiency of packing for the hexagonal close-packed structure.

Table 7-2 lists the unit cell dimensions for some elements that have this structure. The c/a ratios in Table 7-2 suggest that the atoms are not exactly spherical in shape. Only helium has the ideal c/a ratio of 1.633. The ratio is less than the ideal value for all of the other elements in the table, except for Cd and Zn.

TABLE 7-2   UNIT CELL DIMENSIONS OF HEXAGONAL CLOSE-PACKED ELEMENTS

Element	a (Å)	c (Å)	c/a
Be	2.287	3.583	1.567
Cd	2.979	5.618	1.886
Co	2.507	4.069	1.623
Dy	3.590	5.648	1.573
Er	3.559	5.587	1.570
Gd	3.636	5.783	1.590
He	3.57	5.83 (1.45°K)	1.633
Hf	3.197	5.058	1.582
Ho	3.577	5.616	1.570
La	3.75	6.07	1.619
Lu	3.503	5.551	1.585
Mg	3.209	5.210	1.624
Nd	3.657	5.902	1.614
Os	2.735	4.319	1.579
Pr	3.669	5.920	1.614
Re	2.762	4.457	1.614
Ru	2.704	4.282	1.584
Sc	3.309	5.273	1.594
Tb	3.601	5.694	1.581
Ti	2.950	4.686	1.588
Tl	3.456	5.525	1.599
Tm	3.538	5.555	1.570
Y	3.647	5.731	1.571
Zn	2.665	4.947	1.856
Zr	3.232	5.147	1.593

EXERCISE 7-6   Calculate the density of metallic cobalt from the data in
Table 7-2.

7-4   *Body-centered cubic*

The third structure frequently encountered in the elements is body-
centered cubic, for which the space group is *Im3m*. Each atom is
surrounded by eight other atoms at a distance of $2r = (a\sqrt{3}/2)$ (see
Fig. 7-5). There are six next-nearest neighbors at distance a. This is not

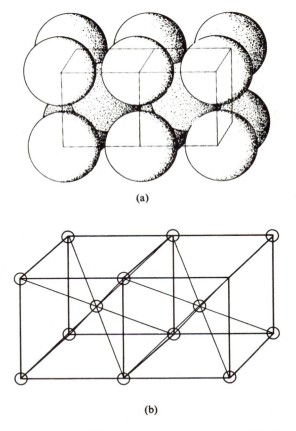

(a)

(b)

FIG. 7-5   *Two unit cells of body-centered cubic structure. (a) Spheres tangent
to each other; (b) point-atom model.*

TABLE 7-3   DIMENSION OF SOME BODY-CENTERED CUBIC
STRUCTURES[a]

Element	(Å)	Element	(Å)
Ba	5.025	Mo	3.147
Cr	2.884	Na	4.291
Cs	6.067 (78° K)	Nb	3.300
Eu	4.606	Rb	5.605 (78° K)
Fe	2.866	Ta	3.306
K	5.247 (78° K)	V	3.024
Li	3.509	W	3.165

[a] Room temperature unless otherwise specified.

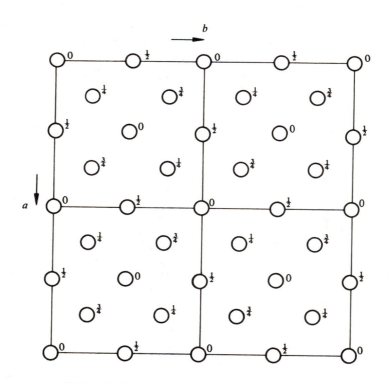

FIG. 7-6   *Four unit cells of the diamond structure.*

a close-packed structure, which would have twelve neighbors equally distributed around each atom, but the combination of eight nearest neighbors and six next-nearest neighbors is favored by several elements (see Table 7-3).

EXERCISE 7-7   Calculate the efficiency of packing in the body-centered cubic structure.

EXERCISE 7-8   Use data from Table 7-3 to calculate (a) the density of niobium metal and (b) the radius of a niobium atom.

7-5   Diamond structure

Four unit cells of the diamond structure are shown in Fig. 7-6. The space group is $Fd3m$, and the atoms occupy the positions $(0,0,0; \frac{1}{4},\frac{1}{4},\frac{1}{4})$ + face centering. The unit cell dimension of diamond is 3.567 Å. This structure is also possessed by silicon ($a = 5.431$ Å), germanium ($a = 5.657$ Å), and gray tin ($a = 6.491$ Å).

EXERCISE 7-9   Calculate the density of diamond.

EXERCISE 7-10   Calculate the efficiency of packing in the diamond structure.

EXERCISE 7-11   (a) On a diagram of the diamond structure (Fig. 7-4), draw lines showing which atoms are bonded to which.

(b) Calculate the length of the C—C bond in diamond.

(c) Calculate the C—C—C bond angle.

EXERCISE 7-12   Calculate the structure factors, in terms of atomic scattering factors f, for the (111), (200), and (220) planes of diamond.

The great strength of diamond crystals is a consequence of the three-dimensional network of strong covalent bonds that link each carbon atom to four other carbon atoms.

7-6   Graphite structure

Polymorphism is quite common among the elements; that is, under different conditions of crystallization different structures result. The diamond structure is actually thermodynamically unstable under ordinary conditions of temperature and pressure, although the rate of

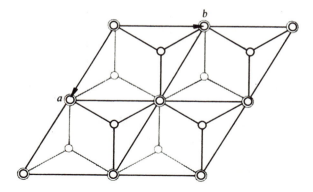

FIG. 7-7 *Four unit cells of the structure of graphite. Network of bonds at z = 0 shown in black; network of bonds at z = ½ shown in gray.*

transition to a more stable form is fortunately very slow. The stable polymorph of carbon is represented by the graphite structure in Fig. 7-7. The hexagonal unit cell has $a = 2.456$ Å, $c = 6.696$ Å. The space group is $P6_3mc$, and there are two atoms in positions $2a:0,0,z; 0,0,\frac{1}{2} + z$, with $z \approx 0$, and two atoms in $2b: \frac{1}{3},\frac{2}{3},z; \frac{2}{3},\frac{1}{3},\frac{1}{2} + z$, with $z \approx 0$. The structure consists of layers in which each atom is bonded to three other atoms to form a hexagonal network. The layers are relatively far apart, which is evidence of only weak bonding between the layers, and this structure accounts for the cleavage and other characteristic properties of graphite. The stacking sequence may be described by translating one layer by $\frac{2}{3},\frac{1}{3},\frac{1}{2}$ with respect to the other.

EXERCISE 7-13   (a) Calculate the density of graphite.

(b) Calculate the length of the C—C bond in a graphite layer. Compare this distance with the values for a carbon-carbon single bond, a double bond and the bond in the benzene molecule.

(c) Calculate the distance between layers.

7-7   Other elements

The few simple structures we have described account for a surprising number of elements. Elements that form polyatomic molecules, such as N_2, O_2, and S_8, necessarily have more complex structures, and each structure is usually unique. A few monatomic elements, such as mercury, have structures that may be regarded as distortions of close-packed structures. Elemental boron has a very complex covalent structure involving icosahedra of boron atoms. Polymorphism is frequently observed in the elements, and iron (which we have listed as body-centered cubic) exhibits several simple phases, whereas four modifications have been observed for manganese.[1]

7-8 *Sodium chloride structure*

The sodium chloride structure was shown in Fig. 1-1, and the calculation of its structure factors was treated in Section 5-22. The space group is $Fm3m$, the Na^+ ions occupy positions (4a): 0,0,0 + face centering, and the Cl^- ions occupy positions (4b): $\frac{1}{2},\frac{1}{2},\frac{1}{2}$ + face centering. Each ion has six neighbors of opposite charge at a distance of $a/2$. There are twelve next-nearest neighbors with like charge at a distance of $a/\sqrt{2}$ along the face diagonals of the cubic unit cell. The unit cell dimension for NaCl is 5.64 Å. Wyckoff[1] lists 220 compounds that have this structure. The electrostatic attractions between ions of opposite charges hold the structure together. The attractions are balanced by the repulsions of ions of the same charge and by a short-range potential due to the finite sizes of the ions. That is, the ions occupy space, and they cannot be pushed arbitrarily close together without generating very strong repulsive forces at short distances. The sizes of the ions evidently are quite important in determining the favorability of this structure. In fact, we can deduce a condition on the ratio of ionic sizes which must be satisfied in order for an RX compound to have the NaCl structure. The ions are in contact along a cell edge, so

$$a = 2[r(Na^+) + r(Cl^-)]$$

Ions of the same type cannot get closer than an ionic diameter. Now, ions of the same type approach each other most closely along the face diagonals of the cell. Assuming that $r(Cl^-) > r(Na^+)$,

$$\frac{a}{\sqrt{2}} \geqslant 2r(Cl^-) \qquad \text{or} \qquad a \geqslant 2\sqrt{2}r(Cl^-)$$

$$2[r(Na^+) + r(Cl^-)] \geqslant 2\sqrt{2}r(Cl^-)$$

$$\frac{r(Cl^-)}{r(Na^+)} \leqslant \sqrt{2} + 1$$

[1] The structures of the elements are summarized by R. W. G. Wyckoff, *Crystal Structures*, 2nd Ed., Vol. 1, Interscience, New York, 1963. This volume also includes the structures of compounds with formulas RX and RX$_2$. The structures of other compounds are treated in other volumes of the Wyckoff series.

A necessary condition for the NaCl structure, is, therefore, that the ratio of ionic sizes be less than 2.414. According to tables of ionic sizes, $r(Cl^-) = 1.81$ Å and $r(Na^+) = 0.96$ Å, so this condition is satisfied for NaCl.

Quantitative calculations of the energy liberated when an ionic crystal, such as NaCl, is formed from infinitely separated ions are given in most textbooks of physical chemistry.[2]

7-9   Cesium chloride structure

This structure was the subject of Exercise 5-13. The space group is $Pm3m$, and there is a Cs^+ ion at $0,0,0$ and a Cl^- ion at $\frac{1}{2},\frac{1}{2},\frac{1}{2}$. One unit cell of the structure is shown in Fig. 7-8. Each ion is surrounded by eight ions of opposite charge at a distance of $a\sqrt{3}/2$, corresponding to one half of the length of the body diagonal of the cube. Each ion has six neighbors of the same charge at a distance a. The radius ratio requirement for this structure is $r(Cl^-)/r(Cs^+) \leqslant (\sqrt{3}+1)/2$. This structure is, therefore, possible when the ions are nearly the same size. If the radius ratio exceeds 1.366, the CsCl structure is not possible, and the NaCl structure is preferred.

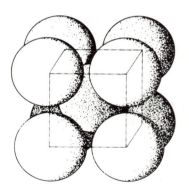

FIG. 7-8   *Structure of CsCl.*

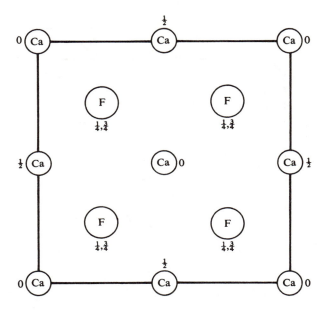

FIG. 7-9   *A unit cell of the CaF$_2$ structure.*

EXERCISE 7-14   Prove that the ratio $r(Cl^-)/r(Cs^+)$ cannot exceed $(\sqrt{3}+1)/2$ for the CsCl structure.

7-10   *Fluorite structure*

Ionic compounds of the type RX_2, where $r(X^-)/r(R^{2+}) \leqslant (\sqrt{3}+1)/2$ may form the CaF$_2$, or fluorite, structure. A unit cell of this structure, shown in Fig. 7-9, has four Ca^{2+} ions at $0,0,0$ + face centering and eight F$^-$ ions at $\frac{1}{4},\frac{1}{4},\frac{1}{4}$; $\frac{3}{4},\frac{3}{4},\frac{3}{4}$ + face centering. The space group is *Fm3m*. The unit cell dimension of CaF$_2$ is 5.462 Å.

EXERCISE 7-15   (a) Determine the number of nearest neighbors of each Ca^{2+} ion in the fluorite structure. What is the Ca^{2+}—F$^-$ distance?

(b) How many nearest neighbors does each F$^-$ ion have?

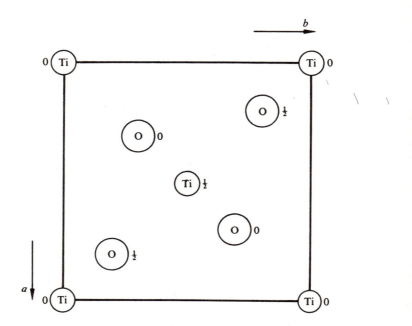

FIG. 7-10   *The tetragonal TiO$_2$ structure projected onto (001). Titanium ions are at 0,0,0; $\frac{1}{2},\frac{1}{2},\frac{1}{2}$. Oxygen ions are at 0.30,0.30,0; 0.80,0.20,$\frac{1}{2}$; 0.70,0.70,0; 0.20,0.80,$\frac{1}{2}$.*

7-11   Rutile structure

The structure possessed by rutile, TiO$_2$, by cassiterite, SnO$_2$, and by a number of other substances with small cations is shown in Fig. 7-10. The structure is tetragonal; for TiO$_2$, $a = 4.594$ Å, $c = 2.958$ Å; for SnO$_2$, $a = 4.737$ Å, $c = 3.186$ Å. The space group is $P4_2/mnm$, the Ti^{4+} ions occupy positions (2a): 0,0,0; $\frac{1}{2},\frac{1}{2},\frac{1}{2}$, and the O^{2-} ions occupy positions (4f): $\pm(x,x,0; \frac{1}{2}+x,\frac{1}{2}-x,\frac{1}{2})$ with x very nearly 0.30. The titanium ion is surrounded by six oxygen ions which form a slightly distorted octahedron.

EXERCISE 7-16   Calculate the distance from the Ti^{4+} ion at $\frac{1}{2},\frac{1}{2},\frac{1}{2}$ in the rutile structure to each of its six O^{2-} neighbors.

EXERCISE 7-17   Describe the nearest neighbor environment of an O^{2-} ion in TiO$_2$. Give the distances wherever necessary.

7-12   Zinc sulfide structure

Zinc blende, ZnS, is cubic. The Zn^{2+} ions are at 0, 0, 0 + face centering, and the S^{2-} ions are at $\frac{1}{4}, \frac{1}{4}, \frac{1}{4}$ + face centering (see Fig. 7-11). The space group is $F\bar{4}3m$, and the lattice dimension for ZnS is 5.409 Å. If the zinc and sulfur atoms were identical, this would be the diamond structure. Each atom in ZnS is surrounded by a regular tetrahedron of atoms of the opposite type.

EXERCISE 7-18   Calculate the structure factors, in terms of f, for the (111), (200), and (220) planes of cubic ZnS. Compare with the corresponding results for diamond from Exercise 7-12.

7-13   Zincite structure

Zincite, ZnO, has a hexagonal structure. The space group is $P6_3mc$, and both types of atom occupy positions $(2b): \frac{1}{3}, \frac{2}{3}, z; \frac{2}{3}, \frac{1}{3}, \frac{1}{2} + z$; with z equal to 0 for zinc and about $\frac{3}{8}$ for oxygen. The unit cell dimensions for ZnO are $a = 3.250$ Å, $c = 5.207$ Å.

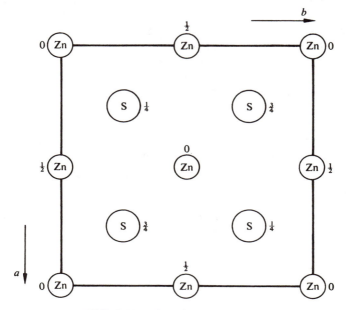

FIG. 7-11   *The cubic ZnS structure.*

As in the case of the zinc blende structure, each atom is surrounded by a tetrahedron of atoms of the opposite type, and both structures consist of continuous networks of interconnected tetrahedra. However, in the case of zincite, symmetry does not require the tetrahedra to be regular. It is instructive to prepare models of these two structures.

EXERCISE 7-19   Draw a diagram of the zincite structure. Select one zinc atom in your diagram and show the four oxygen atoms that surround it. Repeat for an oxygen atom and its four zinc neighbors.

EXERCISE 7-20   Beryllium oxide, BeO, has the zincite structure with $a = 2.698$ Å, $c = 4.380$ Å, and $z = 0.378$. Calculate the distance from a beryllium atom to each of its four neighbors.

7-14   Other structures

The structures described in this chapter represent only a few of the structural types observed in the simple compounds. The examples we have included have been selected because they occur so frequently and because they serve to illustrate structural principles. The crystal structures of molecular compounds are usually all different and are not so readily classified. The reader is referred to structural papers in journals such as *Acta Crystallographica* and *Inorganic Chemistry*, to Wyckoff's compilation of crystal structures,[1] to the summary of structures in *Structure Reports*,[3] and to tabulations of unit cell data such as *Crystal Data*,[4] and Pearson's descriptions of intermetallic structures.[5]

[3] *Structure Reports*, published annually by N. V. A. Oosthoek's Uitgevers Mij, Utrecht, describe the results of crystal structure analyses. Structures published before 1940 are described in *Strukturbericht*.

[4] *Crystal Data*, Determinative Tables (J. D. H. Donnay, G. Donnay, E. G. Cox, O. Kennard, and M. V. King, eds.), Am. Crystallographic Assoc., 1963, tabulates unit cell dimensions for all published structures, other than intermetallic compounds.

[5] W. B. Pearson, *Handbook of Lattice Spacings and Structures of Metals*, Vol. 1, 1958; Vol. 2, 1967, Pergamon, New York.

THE 230 SPACE GROUPS

Point group	Space group	Point group	Space group
Triclinic		Orthorhombic	
1	$P1$		$C222$
$\bar{1}$	$P\bar{1}$		$F222$
			$I222$
Monoclinic			$I2_12_12_1$
2	$P2$	$mm2$	$Pmm2$
	$P2_1$		$Pmc2_1$
	$C2$		$Pcc2$
m	Pm		$Pma2$
	Pc		$Pca2_1$
	Cm		$Pnc2$
	Cc		$Pmn2_1$
$2/m$	$P2/m$		$Pba2$
	$P2_1/m$		$Pna2_1$
	$C2/m$		$Pnn2$
	$P2/c$		$Cmm2$
	$P2_1/c$		$Cmc2_1$
	$C2/c$		$Ccc2$
			$Amm2$
Orthorhombic			$Abm2$
222	$P222$		$Ama2$
	$P222_1$		$Aba2$
	$P2_12_12$		$Fmm2$
	$P2_12_12_1$		$Fdd2$
	$C222_1$		
Orthorhombic (*cont'd.*)		Orthorhombic (*cont'd.*)	

Point group	Space group	Point group	Space group
Orthorhombic		**Tetragonal**	
	$Imm2$		$I4$
	$Iba2$		$I4_1$
	$Ima2$	$\bar{4}$	$P\bar{4}$
mmm	$Pmmm$		$I\bar{4}$
	$Pnnn$	$4/m$	$P4/m$
	$Pccm$		$P4_2/m$
	$Pban$		$P4/n$
	$Pmma$		$P4_2/n$
	$Pnna$		$I4/m$
	$Pmna$		$I4_1/a$
	$Pcca$	422	$P422$
	$Pbam$		$P42_12$
	$Pccn$		$P4_122$
	$Pbcm$		$P4_12_12$
	$Pnnm$		$P4_222$
	$Pmmn$		$P4_22_12$
	$Pbcn$		$P4_322$
	$Pbca$		$P4_32_12$
	$Pnma$		$I422$
	$Cmcm$		$I4_122$
	$Cmca$	$4mm$	$P4mm$
	$Cmmm$		$P4bm$
	$Cccm$		$P4_2cm$
	$Cmma$		$P4_2nm$
	$Ccca$		$P4cc$
	$Fmmm$		$P4nc$
	$Fddd$		$P4_2mc$
	$Immm$		$P4_2bc$
	$Ibam$		$I4mm$
	$Ibca$		$I4cm$
	$Imma$		$I4_1md$
			$I4_1cd$
Tetragonal		$\bar{4}2m$	$P\bar{4}2m$
4	$P4$		$P\bar{4}2c$
	$P4_1$		$P\bar{4}2_1m$
	$P4_2$		$P\bar{4}2_1c$
	$P4_3$		$P\bar{4}m2$
			$P\bar{4}c2$
Tetragonal (*cont'd.*)		**Tetragonal** (*cont'd.*)	

Point group	Space group	Point group	Space group
Tetragonal		Trigonal	
	$P\bar{4}b2$		$P3_121$
	$P\bar{4}n2$		$P3_212$
	$I\bar{4}m2$		$P3_221$
	$I\bar{4}c2$		$R32$
	$I\bar{4}2m$	$3m$	$P3m1$
	$I\bar{4}2d$		$P31m$
$4/mmm$	$P4/mmm$		$P3c1$
	$P4/mcc$		$P31c$
	$P4/nbm$		$R3m$
	$P4/nnc$		$R3c$
	$P4/mbm$	$\bar{3}m$	$P\bar{3}1m$
	$P4/mnc$		$P\bar{3}1c$
	$P4/nmm$		$P\bar{3}m1$
	$P4/ncc$		$P\bar{3}c1$
	$P4_2/mmc$		$R\bar{3}m$
	$P4_2/mcm$		$R\bar{3}c$
	$P4_2/nbc$		
	$P4_2/nnm$	Hexagonal	
	$P4_2/mbc$	6	$P6$
	$P4_2/mnm$		$P6_1$
	$P4_2/nmc$		$P6_5$
	$P4_2/ncm$		$P6_2$
	$I4/mmm$		$P6_4$
	$I4/mcm$		$P6_3$
	$I4_1/amd$	$\bar{6}$	$P\bar{6}$
	$I4_1/acd$	$6/m$	$P6/m$
Trigonal			$P6_3/m$
3	$P3$	622	$P622$
	$P3_1$		$P6_122$
	$P3_2$		$P6_522$
	$R3$		$P6_222$
$\bar{3}$	$P\bar{3}$		$P6_422$
	$R\bar{3}$		$P6_322$
32	$P312$	$6mm$	$P6mm$
	$P321$		$P6cc$
	$P3_112$		$P6_3cm$
Trigonal (*cont'd.*)		Hexagonal (*cont'd.*)	

Point group	Space group		Point group	Space group
Hexagonal		**Cubic**		
$\bar{6}m2$	$P\bar{6}m2$		432	$P432$
	$P\bar{6}c2$			$P4_232$
	$P\bar{6}2m$			$F432$
	$P\bar{6}2c$			$F4_132$
$6/mmm$	$P6/mmm$			$I432$
	$P6/mcc$			$P4_332$
	$P6_3/mcm$			$P4_132$
	$P6_3/mmc$			$I4_132$
			$\bar{4}3m$	$P\bar{4}3m$
				$F\bar{4}3m$
Cubic				$I\bar{4}3m$
23	$P23$			$P\bar{4}3n$
	$F23$			$F\bar{4}3c$
	$I23$			$I\bar{4}3d$
	$P2_13$		$m3m$	$Pm3m$
	$I2_13$			$Pn3n$
$m3$	$Pm3$			$Pm3n$
	$Pn3$			$Pn3m$
	$Fm3$			$Fm3m$
	$Fd3$			$Fm3c$
	$Im3$			$Fd3m$
	$Pa3$			$Fd3c$
	$Ia3$			$Im3m$
Cubic (*cont'd.*)				$Ia3d$

Appendix 2

THE RECIPROCAL
LATTICE

An essential part of the language of crystallography is concerned with the reciprocal lattice. Many seemingly involved geometric calculations become quite simple when considered with the aid of this concept. We will attempt here only a brief description of what it is and why it is useful.

In Chapter 3 we learned how to describe the orientation of planes by means of their Miller indices, and Eq. (3-3) gave us a means of calculating the distance between the members of the set of parallel planes denoted by these indices. It is rather difficult to visualize the orientations of these planes, particularly when the indices may not be small, and when the coordinate system defined by the unit cell edges may be oblique. It is easier to visualize the location of a point in space, so we will define one point for each set of planes as follows. We construct a line from the origin perpendicular to the planes. We place a point at a distance $1/d$ from the origin along this line, where d is the interplanar spacing. This point is the reciprocal lattice point corresponding to the set of planes. The set of all such points, one for each set of parallel planes characterized by a triplet of integral Miller indices, constitutes a lattice.

The advantage of using reciprocal distances may be appreciated by considering Bragg's law in the form

$$\frac{1}{d} = \frac{2 \sin \theta}{\lambda} \qquad \text{(A2-1)}$$

From our films we obtain θ values; the higher the value of θ the lower the corresponding value of d. Our diffraction patterns may thus be regarded as photographs of the reciprocal lattice, distorted by the sine function (the precession camera, in fact, gives an undistorted picture of the reciprocal lattice). The construction in Fig. A2-1, due to P. P. Ewald, shows how this principle can be used to simplify problems in X-ray diffraction. Point O is the origin of the reciprocal lattice. A beam of X rays, of wavelength λ, is passing in the direction AO. We construct a sphere of radius $1/\lambda$ centered on the line AO and passing through the point O. We now keep the sphere fixed and rotate the crystal until

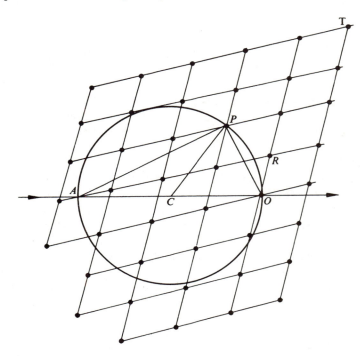

FIG. A2-1 *Ewald construction. Reciprocal lattice centered at O on a sphere of radius $1/\lambda$.*

the reciprocal lattice point P is on the surface of the sphere. The distance OP is $1/d$, by the way we defined the reciprocal lattice. The distance AO is $2/\lambda$, and APO is a right triangle. Therefore,

$$\sin \measuredangle PAO = \frac{OP}{AO} = \frac{1/d}{2/\lambda} = \frac{\lambda}{2d}$$

Therefore, $\measuredangle PAO = \theta$, by Bragg's law, so when the point P is on the surface of the sphere, Bragg's law is satisfied, and the planes corresponding to this point are in reflecting position. By elementary geometry, the angle PCO is 2θ, so CP makes an angle 2θ with the incident X-ray beam, and CP represents the direction of the reflected ray.

If we want to know how to orient the crystal so that the planes represented by reciprocal lattice point R are in reflecting position, we need only consider how we must tilt the crystal so that R is on the surface of the sphere. A point such as T of Fig. A2-1 can never be made to intersect the sphere, so this plane is inaccessible, corresponding to $\sin\theta > 1$. However, we can make the sphere larger by decreasing the wavelength, and point T can be obtained by using shorter wavelengths. (Some writers define the reciprocal lattice distances in terms of λ/d, and use a sphere of unit radius; we prefer to treat the reciprocal lattice as a property of the crystal only.)

The reciprocal lattice concept facilitates many crystallographic calculations. For example, the value of $1/d$ may be computed as the length of the reciprocal lattice vector with coordinates h,k,l referred to a coordinate system based on reciprocal axes. Mastery of this concept is indispensable to crystallographers, but more extensive treatment cannot be given here.

Appendix 3

THE POWDER METHOD

The powder method is somewhat outside the intended scope of this book. However, its widespread use and importance make a short discussion of the principles desirable.

Although the majority of solid substances are crystalline in nature, it is only in rare cases that a solid sample consists of one large crystal. Usually, the sample will be polycrystalline; that is, it is composed of many tiny crystals, and these crystals may have completely random orientations. When such a sample is irradiated with X rays, for every set of planes of the lattice there will be some crystals that are correctly oriented so that Bragg's law is satisfied and reflection will take place. For example, there will be some crystals whose (231) planes make an angle θ with the X-ray beam which is given by

$$\sin \theta = \frac{\lambda}{2d_{231}}$$

If the material is cubic, this becomes

$$\sin \theta = \frac{\lambda \sqrt{14}}{2a}$$

The reflected rays will also be at angle θ with the (231) planes, and the angle between the incident beam and the reflected beam will be 2θ. The locus of all rays that make a particular angle of 2θ with the incident beam is a cone whose half-angle is 2θ. The powdered sample will, therefore, produce a cone for each set of indices, h,k,l, for which reflection is possible. If a photographic film is placed in the path of this radiation, a curved line will be produced for each set of planes of the lattice, and the corresponding θ can be obtained from the position of the line on the film. Alternatively, electronic methods, such as Geiger

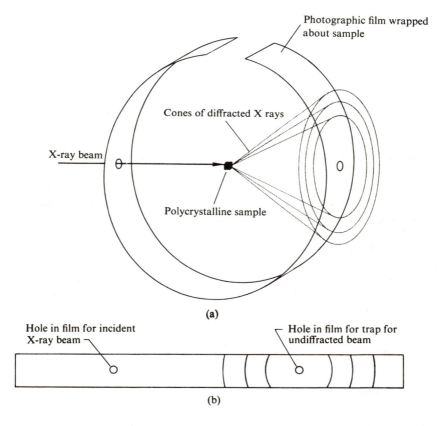

FIG. A3-1 (a) *Production of three powder pattern lines.* (b) *Film after development.*

or proportional counters, can be used to detect and measure the diffracted X rays.

The experimental arrangement is shown in Fig. A3-1a. The photographic film is wrapped cylindrically about the sample. Each crystallite which is correctly oriented for reflection from some set of planes will give rise to a diffracted ray, which will produce a spot if it hits the film. The continuous line which is observed is composed of the spots due to a large number of crystallites. Figure A3-1b shows the film after it has been unwrapped and developed. Only three lines are shown here for simplicity; a typical pattern might have from 10 to 100 lines.

If the crystal structure is cubic, it is possible to analyze completely the powder pattern and to determine the Miller indices of each line and the dimension of the unit cell. For the other crystal systems this complete analysis may not be possible, and the powder pattern cannot give all the information accessible to single crystal methods. However, even in these cases the powder pattern serves to identify the substance. Each crystalline material produces a characteristic powder pattern, and the positions and intensities of the lines can show what a material is and can even serve to identify the components of a mixture.

SOLUTIONS TO EXERCISES

1-1   5.64 Å.

1-2   (a) 210 Å^3; (b) $a = 16.16$ Å, $b = 13.00$ Å, $c = 7.00$ Å, $\alpha = 90°$, $\beta = 90°$, $\gamma = 45.6°$; (c) 1050 Å^3; (d) 5.00, 5 lattice points.

1-3   (a) 2.13 Å; (b) 4.00 Å.

2-4   After filling in all of the obvious products in the table, make use of such relationships as $b(a^2) = (ba)a = (a^2 b)a = a(aba) = ab$, and remember that each element must appear once in each row and column.

2-6   (a) O_h; (b) O_h; (c) T_d.

2-7   Ortho, C_{2v}; meta, C_{2v}; para D_{2h}.

2-8 C_2H_2, $D_{\infty h}$; C_2HCl, $C_{\infty v}$; C_3O_2, $D_{\infty h}$; CH_2CF_2, C_{2v}; *trans*-CHFCHF, C_{2h}; *cis*-CHFCHF, C_{2v}.

2-9 D_3.

2-10 D_{4h}.

2-11 SO_2F_2, C_{2v}; SO_4^{2-}, T_d; $Zn(NH_3)_4^{2+}$, T_d; $CFCl_3$, C_{3v}; CF_2Cl_2, C_{2v}.

2-12   (a) D_{3h}; (b) D_{2h}.

2-13   This is the crown configuration; the structure may be generated by successive applications of symmetry operation S_8. (Note that S_8 is not the same group as D_{4d}, however, since S_8 does not generate the diagonal mirror planes.)

2-14 $CrCl_6^{3-}$, O_h; $CrCl_5Br^{3-}$, C_{4v}; *trans*-$CrCl_4Br_2^{3-}$, D_{4h}; *cis*-$CrCl_4Br_2^{3-}$, C_{2v}; one isomer of $CrCl_3Br_3^{3-}$ has C_{3v} symmetry, the other has C_{2v} symmetry.

2-15 (a) planar; (b) trigonal pyramid.

2-16 (a) $D_{\infty h}$; (b) D_{2h}; (c) C_{4v}; (d) D_{3h}.

3-1 Lattice points at $0,\frac{1}{2},\frac{1}{2}$ and $\frac{1}{2},\frac{1}{2},0$ imply a lattice point at $\frac{1}{2},0,\frac{1}{2}$, so the B face would also have to be centered.

3-2 Choose a primitive monoclinic cell with vectors to the points $\frac{1}{2},0,\frac{1}{2}$ and $\frac{1}{2},0,-\frac{1}{2}$.

3-3 If points $\frac{1}{2},\frac{1}{2},\frac{1}{2}$ and $\frac{1}{2},0,\frac{1}{2}$ are lattice points, then $0,\frac{1}{2},0$ must also be a lattice point. Similarly, $\frac{1}{2},0,0$ and $0,0,\frac{1}{2}$ must be lattice points, and a primitive orthorhombic cell can be chosen with axes $a/2$, $b/2$, and $c/2$.

3-4 Tetragonal symmetry would require B centering as well as A centering. Both A and B centering implies face centering (see Exercise 3-1), and a body-centered tetragonal lattice could be chosen.

3-5 (a) 114.13 Å^3; (b) $a = 6.09$ Å; $c = 10.67$ Å; (c) 342.4 Å^3.

3-6 $a = 6.00\sqrt{2} = 8.48$ Å. The classification is based on symmetry. A cubic crystal has four threefold axes, whereas a rhombohedral crystal has one axis of threefold symmetry.

3-7 (a) 432; (b) 025; (c) 3 0 12; (d) 650; (e) 650.

3-10 3.27 Å.

3-11 4.88 Å.

4-1 Pd(1) has four S neighbors at 2.396 Å; Pd(2) has four S neighbors at 2.267 Å; Pd(3) has two S neighbors at 2.447 Å and two at 2.233 Å. The four neighbors of a sulfur atom are Pd(1) at 2.396 Å, Pd(2) at 2.267 Å, Pd(3) at 2.447 Å, and Pd(4) at 2.233 Å.

4-2 Hg—Br(1) = 2.504 Å, Hg—Br(2) = 2.499 Å, angle = 180°.

4-3 General positions: x,y,z; $\bar{x},y,z$; $x,\bar{y},\frac{1}{2}+z$; $\bar{x},\bar{y},\frac{1}{2}+z$. Special positions: $0,y,z$; $0,\bar{y},\frac{1}{2}+z$. Special positions: $\frac{1}{2},y,z$; $\frac{1}{2},\bar{y},\frac{1}{2}+z$.

4-4 General positions: x,y,z; $\bar{y},x-y,z$; $y-x,\bar{x},z$; $x,x-y,\bar{z}$; $y-x,y,\bar{z}$; $\bar{y},\bar{x},\bar{z}$. Special positions: $0,0,z$; $0,0,\bar{z}$. There are also threefold axes parallel to c through the points $x=\frac{2}{3}$, $y=\frac{1}{3}$ and $x=\frac{1}{3}$, $y=\frac{2}{3}$.

4-5 Two at 1.606 Å, two at 1.599 Å.

4-6 Bond length = 1.971 Å. Each Cl atom has two neighbors at 3.34 Å, two at 3.72 Å, four at 3.84 Å, and two at 3.98 Å.

4-7 (b) Si, C_{3v}; O(1), D_{3h}; O(2), C_{2h}; (c) one O(1) at 1.562 Å, three O(3) at 1.534 Å.

4-8 By suitable shifts of origin, both sets of positions can be written C centering $\pm (x,y,z; \frac{1}{2}-x,y,\frac{1}{2}-z)$.

5-1 $n = 1$, 72.5°; $n = 2$, 84.3°.

5-2 $\alpha_0 = 10°$, $\Delta\alpha = 0.013°$; $\alpha_0 = 1°$, $\Delta\alpha = 0.124°$; $\alpha_0 = 0.1°$, $\Delta\alpha = 0.422°$.

5-4 Average 6.05 Å.

5-5 (a) 5.50°; (b) 11.04°; (c) 32.0°; (d) 27.7°.

5-6 (a) 8.14 Å; (b) $a = 6.50$ Å; $c = 8.49$ Å; $\beta = 94.9°$; (c) $V = 448$ Å^3, density = 2.35 g/ml.

5-7 For 112, $\theta = 17.77°$; for all others, $\theta = 12.89°$.

5-8   (a) $(-1)^n$; (b) $-i$; (c) i; (d) Take logarithm of $-1 = \exp[\pi i(2n + 1)]$, $\ln(-1) = \pi i(2n + 1)$.

5-10   (a) x,y,z; $\bar{y},x,z + \frac{1}{4}$; $\bar{x},\bar{y},z + \frac{1}{2}$; $y,\bar{x},z + \frac{3}{4}$; (b) Show that these structure factors differ only by a constant complex factor. For example, $F(\bar{k}hl) = e^{2\pi i l/4} F(hkl)$, so these planes will give equal intensities.

5-11   Write formulas for these structure factors and observe that they differ at most by the sign of the exponent.

5-13   All reflections present. The structure factor is the sum of the atomic scattering factors when $h + k + l$ is even, and the difference when $h + k + l$ is odd. This lattice is primitive. The CsCl structure is primitive cubic, contrary to the opinion of many textbook writers.

5-15   (a) $hkl, h + k = 2n$; $h0l, l = 2n$; (b) $hkl, k + l = 2n$; $0kl, k = 2n$; $h0l, h = 2n$; (c) $hkl, h + k + l = 2n$; $hk0, h = 2n$.

5-16   (a) This centering violates the sixfold symmetry required for the hexagonal system. (b) $a = b = 8.77$ Å, $c = 7.17$ Å, $\alpha = \beta = 90°$, $\gamma = 89.8°$. (c) In terms of these "tetragonal" lattice vectors, a $hk0$ Weissenberg photograph or precession photograph would provide a useful indication of the true symmetry. Complete verification of tetragonal symmetry would require additional pictures, but coesite actually is monoclinic.

5-17   (a) Reflections missing unless $-h + k + l = 4n$; (b) this distribution of lattice points lacks tetragonal symmetry; (c) $a = 22.58$ Å, $b = 22.58$ Å, $c = 12.03$ Å, $\beta = 118.0°$.

6-2   A possible set is $0.00, 0.00$; $0.30, 0.31$; $0.85, 0.52$. Equally valid structures are obtained by changing the signs of all of these coordinates or by any shift of origin.

7-2   2.972 g/ml.

7-3 P indicates a primitive lattice; 6_3 indicates a sixfold screw axis parallel to c; first m indicates a mirror plane normal to c; second m indicates mirror planes normal to a and b (and symmetry-related directions); c represents glide planes with glide component $c/2$ and orientation including the c and a axes (and symmetry-related directions).

7-5   Same as cubic close-packed; 74.05 %.

7-6   8.84 g/ml.

7-7 $\sqrt{3}\pi/8 = 0.6802$.

7-8   (a) 8.59 g/ml; (b) 1.429 Å.

7-9   3.515 g/ml.

7-10 $\sqrt{3}\pi/16 = 0.3401$.

7-11   (b) 1.545 Å; (c) 109.47°.

7-12 $F(111) = 4f(1 - i)$, $F(200) = 0$, $F(220) = 8f$.

7-13   (a) 2.281 g/ml; (b) graphite, 1.418 Å; single bond, 1.544 Å; double bond, 1.334 Å; triple bond, 1.206 Å; benzene, 1.395 Å; (c) 3.348 Å.

7-15 (a) Eight neighbors at 2.366 Å; (b) four neighbors.

7-16 Two at 1.95 Å, four at 1.97 Å.

7-17 Two Ti at 1.97 Å, one Ti at 1.95 Å, one O at 2.60 Å.

7-18 $F(111) = 4(f_{Zn} - if_S)$, $F(200) = 4(f_{Zn} - f_S)$, $F(220) = 4(f_{Zn} + f_S)$.

7-20 One oxygen neighbor at 1.642 Å, three at 1.651 Å.

INDEX

A CATALOG OF SELECTED
DOVER BOOKS
IN SCIENCE AND MATHEMATICS

A CATALOG OF SELECTED
DOVER BOOKS
IN SCIENCE AND MATHEMATICS

QUALITATIVE THEORY OF DIFFERENTIAL EQUATIONS, V.V. Nemytskii and V.V. Stepanov. Classic graduate-level text by two prominent Soviet mathematicians covers classical differential equations as well as topological dynamics and ergodic theory. Bibliographies. 523pp. 5⅜ x 8½. 65954-2 Pa. $14.95

MATRICES AND LINEAR ALGEBRA, Hans Schneider and George Phillip Barker. Basic textbook covers theory of matrices and its applications to systems of linear equations and related topics such as determinants, eigenvalues and differential equations. Numerous exercises. 432pp. 5⅜ x 8½. 66014-1 Pa. $10.95

QUANTUM THEORY, David Bohm. This advanced undergraduate-level text presents the quantum theory in terms of qualitative and imaginative concepts, followed by specific applications worked out in mathematical detail. Preface. Index. 655pp. 5⅜ x 8½. 65969-0 Pa. $14.95

ATOMIC PHYSICS (8th edition), Max Born. Nobel laureate's lucid treatment of kinetic theory of gases, elementary particles, nuclear atom, wave-corpuscles, atomic structure and spectral lines, much more. Over 40 appendices, bibliography. 495pp. 5⅜ x 8½. 65984-4 Pa. $13.95

ELECTRONIC STRUCTURE AND THE PROPERTIES OF SOLIDS: The Physics of the Chemical Bond, Walter A. Harrison. Innovative text offers basic understanding of the electronic structure of covalent and ionic solids, simple metals, transition metals and their compounds. Problems. 1980 edition. 582pp. 6⅛ x 9¼. 66021-4 Pa. $16.95

BOUNDARY VALUE PROBLEMS OF HEAT CONDUCTION, M. Necati Özisik. Systematic, comprehensive treatment of modern mathematical methods of solving problems in heat conduction and diffusion. Numerous examples and problems. Selected references. Appendices. 505pp. 5⅜ x 8½. 65990-9 Pa. $12.95

A SHORT HISTORY OF CHEMISTRY (3rd edition), J.R. Partington. Classic exposition explores origins of chemistry, alchemy, early medical chemistry, nature of atmosphere, theory of valency, laws and structure of atomic theory, much more. 428pp. 5⅜ x 8½. (Available in U.S. only) 65977-1 Pa. $11.95

A HISTORY OF ASTRONOMY, A. Pannekoek. Well-balanced, carefully reasoned study covers such topics as Ptolemaic theory, work of Copernicus, Kepler, Newton, Eddington's work on stars, much more. Illustrated. References. 521pp. 5⅜ x 8½. 65994-1 Pa. $12.95

PRINCIPLES OF METEOROLOGICAL ANALYSIS, Walter J. Saucier. Highly respected, abundantly illustrated classic reviews atmospheric variables, hydrostatics, static stability, various analyses (scalar, cross-section, isobaric, isentropic, more). For intermediate meteorology students. 454pp. 6⅛ x 9¼. 65979-8 Pa. $14.95

THE ELECTROMAGNETIC FIELD, Albert Shadowitz. Comprehensive undergraduate text covers basics of electric and magnetic fields, builds up to electromagnetic theory. Also related topics, including relativity. Over 900 problems. 768pp. 5⅝ x 8¼. 65660-8 Pa. $18.95

FOURIER SERIES, Georgi P. Tolstov. Translated by Richard A. Silverman. A valuable addition to the literature on the subject, moving clearly from subject to subject and theorem to theorem. 107 problems, answers. 336pp. 5⅜ x 8½. 63317-9 Pa. $9.95

THEORY OF ELECTROMAGNETIC WAVE PROPAGATION, Charles Herach Papas. Graduate-level study discusses the Maxwell field equations, radiation from wire antennas, the Doppler effect and more. xiii + 244pp. 5⅜ x 8½. 65678-0 Pa. $6.95

DISTRIBUTION THEORY AND TRANSFORM ANALYSIS: An Introduction to Generalized Functions, with Applications, A.H. Zemanian. Provides basics of distribution theory, describes generalized Fourier and Laplace transformations. Numerous problems. 384pp. 5⅜ x 8½. 65479-6 Pa. $11.95

THE PHYSICS OF WAVES, William C. Elmore and Mark A. Heald. Unique overview of classical wave theory. Acoustics, optics, electromagnetic radiation, more. Ideal as classroom text or for self-study. Problems. 477pp. 5⅜ x 8½. 64926-1 Pa. $13.95

CALCULUS OF VARIATIONS WITH APPLICATIONS, George M. Ewing. Applications-oriented introduction to variational theory develops insight and promotes understanding of specialized books, research papers. Suitable for advanced undergraduate/graduate students as primary, supplementary text. 352pp. 5⅜ x 8½. 64856-7 Pa. $9.95

A TREATISE ON ELECTRICITY AND MAGNETISM, James Clerk Maxwell. Important foundation work of modern physics. Brings to final form Maxwell's theory of electromagnetism and rigorously derives his general equations of field theory. 1,084pp. 5⅜ x 8½. 60636-8, 60637-6 Pa., Two-vol. set $25.90

AN INTRODUCTION TO THE CALCULUS OF VARIATIONS, Charles Fox. Graduate-level text covers variations of an integral, isoperimetrical problems, least action, special relativity, approximations, more. References. 279pp. 5⅜ x 8½. 65499-0 Pa. $8.95

HYDRODYNAMIC AND HYDROMAGNETIC STABILITY, S. Chandrasekhar. Lucid examination of the Rayleigh-Benard problem; clear coverage of the theory of instabilities causing convection. 704pp. 5⅝ x 8¼. 64071-X Pa. $14.95

CALCULUS OF VARIATIONS, Robert Weinstock. Basic introduction covering isoperimetric problems, theory of elasticity, quantum mechanics, electrostatics, etc. Exercises throughout. 326pp. 5⅜ x 8½. 63069-2 Pa. $9.95

DYNAMICS OF FLUIDS IN POROUS MEDIA, Jacob Bear. For advanced students of ground water hydrology, soil mechanics and physics, drainage and irrigation engineering and more. 335 illustrations. Exercises, with answers. 784pp. 6⅛ x 9¼. 65675-6 Pa. $19.95

CATALOG OF DOVER BOOKS

NUMERICAL METHODS FOR SCIENTISTS AND ENGINEERS, Richard Hamming. Classic text stresses frequency approach in coverage of algorithms, polynomial approximation, Fourier approximation, exponential approximation, other topics. Revised and enlarged 2nd edition. 721pp. 5⅜ x 8½. 65241-6 Pa. $15.95

THEORETICAL SOLID STATE PHYSICS, Vol. 1: Perfect Lattices in Equilibrium; Vol. II: Non-Equilibrium and Disorder, William Jones and Norman H. March. Monumental reference work covers fundamental theory of equilibrium properties of perfect crystalline solids, non-equilibrium properties, defects and disordered systems. Appendices. Problems. Preface. Diagrams. Index. Bibliography. Total of 1,301pp. 5⅜ x 8½. Two volumes.
Vol. I: 65015-4 Pa. $16.95
Vol. II: 65016-2 Pa. $16.95

OPTIMIZATION THEORY WITH APPLICATIONS, Donald A. Pierre. Broad spectrum approach to important topic. Classical theory of minima and maxima, calculus of variations, simplex technique and linear programming, more. Many problems, examples. 640pp. 5⅜ x 8½. 65205-X Pa. $16.95

THE CONTINUUM: A Critical Examination of the Foundation of Analysis, Hermann Weyl. Classic of 20th-century foundational research deals with the conceptual problem posed by the continuum. 156pp. 5⅜ x 8½. 67982-9 Pa. $6.95

ESSAYS ON THE THEORY OF NUMBERS, Richard Dedekind. Two classic essays by great German mathematician: on the theory of irrational numbers; and on transfinite numbers and properties of natural numbers. 115pp. 5⅜ x 8½. 21010-3 Pa. $5.95

THE FUNCTIONS OF MATHEMATICAL PHYSICS, Harry Hochstadt. Comprehensive treatment of orthogonal polynomials, hypergeometric functions, Hill's equation, much more. Bibliography. Index. 322pp. 5⅜ x 8½. 65214-9 Pa. $9.95

NUMBER THEORY AND ITS HISTORY, Oystein Ore. Unusually clear, accessible introduction covers counting, properties of numbers, prime numbers, much more. Bibliography. 380pp. 5⅜ x 8½. 65620-9 Pa. $10.95

THE VARIATIONAL PRINCIPLES OF MECHANICS, Cornelius Lanczos. Graduate level coverage of calculus of variations, equations of motion, relativistic mechanics, more. First inexpensive paperbound edition of classic treatise. Index. Bibliography. 418pp. 5⅜ x 8½. 65067-7 Pa. $12.95

MATHEMATICAL TABLES AND FORMULAS, Robert D. Carmichael and Edwin R. Smith. Logarithms, sines, tangents, trig functions, powers, roots, reciprocals, exponential and hyperbolic functions, formulas and theorems. 269pp. 5⅜ x 8½. 60111-0 Pa. $6.95

THEORETICAL PHYSICS, Georg Joos, with Ira M. Freeman. Classic overview covers essential math, mechanics, electromagnetic theory, thermodynamics, quantum mechanics, nuclear physics, other topics. First paperback edition. xxiii + 885pp. 5⅜ x 8½. 65227-0 Pa. $21.95

CATALOG OF DOVER BOOKS

HANDBOOK OF MATHEMATICAL FUNCTIONS WITH FORMULAS, GRAPHS, AND MATHEMATICAL TABLES, edited by Milton Abramowitz and Irene A. Stegun. Vast compendium: 29 sets of tables, some to as high as 20 places. 1,046pp. 8 x 10½. 61272-4 Pa. $26.95

MATHEMATICAL METHODS IN PHYSICS AND ENGINEERING, John W. Dettman. Algebraically based approach to vectors, mapping, diffraction, other topics in applied math. Also generalized functions, analytic function theory, more. Exercises. 448pp. 5⅜ x 8¼. 65649-7 Pa. $10.95

A SURVEY OF NUMERICAL MATHEMATICS, David M. Young and Robert Todd Gregory. Broad self-contained coverage of computer-oriented numerical algorithms for solving various types of mathematical problems in linear algebra, ordinary and partial, differential equations, much more. Exercises. Total of 1,248pp. 5⅜ x 8¼. Two volumes. Vol. I: 65691-8 Pa. $16.95
Vol. II: 65692-6 Pa. $16.95

TENSOR ANALYSIS FOR PHYSICISTS, J.A. Schouten. Concise exposition of the mathematical basis of tensor analysis, integrated with well-chosen physical examples of the theory. Exercises. Index. Bibliography. 289pp. 5⅜ x 8½. 65582-2 Pa. $8.95

INTRODUCTION TO NUMERICAL ANALYSIS (2nd Edition), F.B. Hildebrand. Classic, fundamental treatment covers computation, approximation, interpolation, numerical differentiation and integration, other topics. 150 new problems. 669pp. 5⅜ x 8½. 65363-3 Pa. $16.95

INVESTIGATIONS ON THE THEORY OF THE BROWNIAN MOVEMENT, Albert Einstein. Five papers (1905–8) investigating dynamics of Brownian motion and evolving elementary theory. Notes by R. Fürth. 122pp. 5⅜ x 8½. 60304-0 Pa. $5.95

CATASTROPHE THEORY FOR SCIENTISTS AND ENGINEERS, Robert Gilmore. Advanced-level treatment describes mathematics of theory grounded in the work of Poincaré, R. Thom, other mathematicians. Also important applications to problems in mathematics, physics, chemistry and engineering. 1981 edition. References. 28 tables. 397 black-and-white illustrations. xvii + 666pp. 6⅛ x 9¼. 67539-4 Pa. $17.95

AN INTRODUCTION TO STATISTICAL THERMODYNAMICS, Terrell L. Hill. Excellent basic text offers wide-ranging coverage of quantum statistical mechanics, systems of interacting molecules, quantum statistics, more. 523pp. 5⅜ x 8½. 65242-4 Pa. $12.95

STATISTICAL PHYSICS, Gregory H. Wannier. Classic text combines thermodynamics, statistical mechanics and kinetic theory in one unified presentation of thermal physics. Problems with solutions. Bibliography. 532pp. 5⅜ x 8½. 65401-X Pa. $12.95

ORDINARY DIFFERENTIAL EQUATIONS, Morris Tenenbaum and Harry Pollard. Exhaustive survey of ordinary differential equations for undergraduates in mathematics, engineering, science. Thorough analysis of theorems. Diagrams. Bibliography. Index. 818pp. 5⅜ x 8½. 64940-7 Pa. $18.95

STATISTICAL MECHANICS: Principles and Applications, Terrell L. Hill. Standard text covers fundamentals of statistical mechanics, applications to fluctuation theory, imperfect gases, distribution functions, more. 448pp. 5⅜ x 8½. 65390-0 Pa. $11.95

ORDINARY DIFFERENTIAL EQUATIONS AND STABILITY THEORY: An Introduction, David A. Sánchez. Brief, modern treatment. Linear equation, stability theory for autonomous and nonautonomous systems, etc. 164pp. 5⅜ x 8¼. 63828-6 Pa. $6.95

THIRTY YEARS THAT SHOOK PHYSICS: The Story of Quantum Theory, George Gamow. Lucid, accessible introduction to influential theory of energy and matter. Careful explanations of Dirac's anti-particles, Bohr's model of the atom, much more. 12 plates. Numerous drawings. 240pp. 5⅜ x 8½. 24895-X Pa. $7.95

THEORY OF MATRICES, Sam Perlis. Outstanding text covering rank, nonsingularity and inverses in connection with the development of canonical matrices under the relation of equivalence, and without the intervention of determinants. Includes exercises. 237pp. 5⅜ x 8½. 66810-X Pa. $8.95

GREAT EXPERIMENTS IN PHYSICS: Firsthand Accounts from Galileo to Einstein, edited by Morris H. Shamos. 25 crucial discoveries: Newton's laws of motion, Chadwick's study of the neutron, Hertz on electromagnetic waves, more. Original accounts clearly annotated. 370pp. 5⅜ x 8½. 25346-5 Pa. $10.95

INTRODUCTION TO PARTIAL DIFFERENTIAL EQUATIONS WITH APPLICATIONS, E.C. Zachmanoglou and Dale W. Thoe. Essentials of partial differential equations applied to common problems in engineering and the physical sciences. Problems and answers. 416pp. 5⅜ x 8½. 65251-3 Pa. $11.95

BURNHAM'S CELESTIAL HANDBOOK, Robert Burnham, Jr. Thorough guide to the stars beyond our solar system. Exhaustive treatment. Alphabetical by constellation: Andromeda to Cetus in Vol. 1; Chamaeleon to Orion in Vol. 2; and Pavo to Vulpecula in Vol. 3. Hundreds of illustrations. Index in Vol. 3. 2,000pp. 6⅛ x 9¼. 23567-X, 23568-8, 23673-0 Pa., Three-vol. set $44.85

CHEMICAL MAGIC, Leonard A. Ford. Second Edition, Revised by E. Winston Grundmeier. Over 100 unusual stunts demonstrating cold fire, dust explosions, much more. Text explains scientific principles and stresses safety precautions. 128pp. 5⅜ x 8½. 67628-5 Pa. $5.95

AMATEUR ASTRONOMER'S HANDBOOK, J.B. Sidgwick. Timeless, comprehensive coverage of telescopes, mirrors, lenses, mountings, telescope drives, micrometers, spectroscopes, more. 189 illustrations. 576pp. 5⅜ x 8¼. (Available in U.S. only) 24034-7 Pa. $11.95

SPECIAL FUNCTIONS, N.N. Lebedev. Translated by Richard Silverman. Famous Russian work treating more important special functions, with applications to specific problems of physics and engineering. 38 figures. 308pp. 5⅜ x 8½. 60624-4 Pa. $9.95

OBSERVATIONAL ASTRONOMY FOR AMATEURS, J.B. Sidgwick. Mine of useful data for observation of sun, moon, planets, asteroids, aurorae, meteors, comets, variables, binaries, etc. 39 illustrations. 384pp. 5⅜ x 8¼. (Available in U.S. only) 24033-9 Pa. $8.95

INTEGRAL EQUATIONS, F.G. Tricomi. Authoritative, well-written treatment of extremely useful mathematical tool with wide applications. Volterra Equations, Fredholm Equations, much more. Advanced undergraduate to graduate level. Exercises. Bibliography. 238pp. 5⅜ x 8½. 64828-1 Pa. $8.95

POPULAR LECTURES ON MATHEMATICAL LOGIC, Hao Wang. Noted logician's lucid treatment of historical developments, set theory, model theory, recursion theory and constructivism, proof theory, more. 3 appendixes. Bibliography. 1981 edition. ix + 283pp. 5⅜ x 8½. 67632-3 Pa. $8.95

MODERN NONLINEAR EQUATIONS, Thomas L. Saaty. Emphasizes practical solution of problems; covers seven types of equations. ". . . a welcome contribution to the existing literature...."–Math Reviews. 490pp. 5⅜ x 8½. 64232-1 Pa. $13.95

FUNDAMENTALS OF ASTRODYNAMICS, Roger Bate et al. Modern approach developed by U.S. Air Force Academy. Designed as a first course. Problems, exercises. Numerous illustrations. 455pp. 5⅜ x 8½. 60061-0 Pa. $10.95

INTRODUCTION TO LINEAR ALGEBRA AND DIFFERENTIAL EQUATIONS, John W. Dettman. Excellent text covers complex numbers, determinants, orthonormal bases, Laplace transforms, much more. Exercises with solutions. Undergraduate level. 416pp. 5⅜ x 8½. 65191-6 Pa. $11.95

INCOMPRESSIBLE AERODYNAMICS, edited by Bryan Thwaites. Covers theoretical and experimental treatment of the uniform flow of air and viscous fluids past two-dimensional aerofoils and three-dimensional wings; many other topics. 654pp. 5⅜ x 8½. 65465-6 Pa. $16.95

INTRODUCTION TO DIFFERENCE EQUATIONS, Samuel Goldberg. Exceptionally clear exposition of important discipline with applications to sociology, psychology, economics. Many illustrative examples; over 250 problems. 260pp. 5⅜ x 8½. 65084-7 Pa. $8.95

LAMINAR BOUNDARY LAYERS, edited by L. Rosenhead. Engineering classic covers steady boundary layers in two- and three- dimensional flow, unsteady boundary layers, stability, observational techniques, much more. 708pp. 5⅜ x 8½. 65646-2 Pa. $18 95

LECTURES ON CLASSICAL DIFFERENTIAL GEOMETRY, Second Edition, Dirk J. Struik. Excellent brief introduction covers curves, theory of surfaces, fundamental equations, geometry on a surface, conformal mapping, other topics. Problems. 240pp. 5⅜ x 8½. 65609-8 Pa. $8.95

ROTARY-WING AERODYNAMICS, W.Z. Stepniewski. Clear, concise text covers aerodynamic phenomena of the rotor and offers guidelines for helicopter performance evaluation. Originally prepared for NASA. 537 figures. 640pp. 6⅛ x 9¼.
64647-5 Pa. $16.95

DIFFERENTIAL GEOMETRY, Heinrich W. Guggenheimer. Local differential geometry as an application of advanced calculus and linear algebra. Curvature, transformation groups, surfaces, more. Exercises. 62 figures. 378pp. 5⅜ x 8½.
63433-7 Pa. $9.95

INTRODUCTION TO SPACE DYNAMICS, William Tyrrell Thomson. Comprehensive, classic introduction to space-flight engineering for advanced undergraduate and graduate students. Includes vector algebra, kinematics, transformation of coordinates. Bibliography. Index. 352pp. 5⅜ x 8½.
65113-4 Pa. $9.95

A SURVEY OF MINIMAL SURFACES, Robert Osserman. Up-to-date, in-depth discussion of the field for advanced students. Corrected and enlarged edition covers new developments. Includes numerous problems. 192pp. 5⅜ x 8½.
64998-9 Pa. $8.95

ANALYTICAL MECHANICS OF GEARS, Earle Buckingham. Indispensable reference for modern gear manufacture covers conjugate gear-tooth action, gear-tooth profiles of various gears, many other topics. 263 figures. 102 tables. 546pp. 5⅜ x 8½.
65712-4 Pa. $14.95

SET THEORY AND LOGIC, Robert R. Stoll. Lucid introduction to unified theory of mathematical concepts. Set theory and logic seen as tools for conceptual understanding of real number system. 496pp. 5⅜ x 8¼.
63829-4 Pa. $12.95

A HISTORY OF MECHANICS, René Dugas. Monumental study of mechanical principles from antiquity to quantum mechanics. Contributions of ancient Greeks, Galileo, Leonardo, Kepler, Lagrange, many others. 671pp. 5⅜ x 8½.
65632-2 Pa. $14.95

FAMOUS PROBLEMS OF GEOMETRY AND HOW TO SOLVE THEM, Benjamin Bold. Squaring the circle, trisecting the angle, duplicating the cube: learn their history, why they are impossible to solve, then solve them yourself. 128pp. 5⅜ x 8½.
24297-8 Pa. $4.95

MECHANICAL VIBRATIONS, J.P. Den Hartog. Classic textbook offers lucid explanations and illustrative models, applying theories of vibrations to a variety of practical industrial engineering problems. Numerous figures. 233 problems, solutions. Appendix. Index. Preface. 436pp. 5⅜ x 8½.
64785-4 Pa. $11.95

CURVATURE AND HOMOLOGY, Samuel I. Goldberg. Thorough treatment of specialized branch of differential geometry. Covers Riemannian manifolds, topology of differentiable manifolds, compact Lie groups, other topics. Exercises. 315pp. 5⅜ x 8½.
64314-X Pa. $9.95

HISTORY OF STRENGTH OF MATERIALS, Stephen P. Timoshenko. Excellent historical survey of the strength of materials with many references to the theories of elasticity and structure. 245 figures. 452pp. 5⅜ x 8½.
61187-6 Pa. $12.95

CATALOG OF DOVER BOOKS

GEOMETRY OF COMPLEX NUMBERS, Hans Schwerdtfeger. Illuminating, widely praised book on analytic geometry of circles, the Moebius transformation, and two-dimensional non-Euclidean geometries. 200pp. 5⅜ x 8¼. 63830-8 Pa. $8.95

MECHANICS, J.P. Den Hartog. A classic introductory text or refresher. Hundreds of applications and design problems illuminate fundamentals of trusses, loaded beams and cables, etc. 334 answered problems. 462pp. 5⅜ x 8½. 60754-2 Pa. $11.95

TOPOLOGY, John G. Hocking and Gail S. Young. Superb one-year course in classical topology. Topological spaces and functions, point-set topology, much more. Examples and problems. Bibliography. Index. 384pp. 5⅜ x 8¼. 65676-4 Pa. $10.95

STRENGTH OF MATERIALS, J.P. Den Hartog. Full, clear treatment of basic material (tension, torsion, bending, etc.) plus advanced material on engineering methods, applications. 350 answered problems. 323pp. 5⅜ x 8½. 60755-0 Pa. $9.95

ELEMENTARY CONCEPTS OF TOPOLOGY, Paul Alexandroff. Elegant, intuitive approach to topology from set-theoretic topology to Betti groups; how concepts of topology are useful in math and physics. 25 figures. 57pp. 5⅜ x 8½.
60747-X Pa. $3.95

ADVANCED STRENGTH OF MATERIALS, J.P. Den Hartog. Superbly written advanced text covers torsion, rotating disks, membrane stresses in shells, much more. Many problems and answers. 388pp. 5⅜ x 8½. 65407-9 Pa. $10.95

COMPUTABILITY AND UNSOLVABILITY, Martin Davis. Classic graduate-level introduction to theory of computability, usually referred to as theory of recurrent functions. New preface and appendix. 288pp. 5⅜ x 8½. 61471-9 Pa. $8.95

GENERAL CHEMISTRY, Linus Pauling. Revised 3rd edition of classic first-year text by Nobel laureate. Atomic and molecular structure, quantum mechanics, statistical mechanics, thermodynamics correlated with descriptive chemistry. Problems. 992pp. 5⅜ x 8½. 65622-5 Pa. $19.95

AN INTRODUCTION TO MATRICES, SETS AND GROUPS FOR SCIENCE STUDENTS, G. Stephenson. Concise, readable text introduces sets, groups, and most importantly, matrices to undergraduate students of physics, chemistry, and engineering. Problems. 164pp. 5⅜ x 8½. 65077-4 Pa. $7.95

THE HISTORICAL BACKGROUND OF CHEMISTRY, Henry M. Leicester. Evolution of ideas, not individual biography. Concentrates on formulation of a coherent set of chemical laws. 260pp. 5⅜ x 8½. 61053-5 Pa. $8.95

THE PHILOSOPHY OF MATHEMATICS: An Introductory Essay, Stephan Körner. Surveys the views of Plato, Aristotle, Leibniz & Kant concerning propositions and theories of applied and pure mathematics. Introduction. Two appendices. Index. 198pp. 5⅜ x 8½. 25048-2 Pa. $8.95

THE DEVELOPMENT OF MODERN CHEMISTRY, Aaron J. Ihde. Authoritative history of chemistry from ancient Greek theory to 20th-century innovation. Covers major chemists and their discoveries. 209 illustrations. 14 tables. Bibliographies. Indices. Appendices. 851pp. 5⅜ x 8½. 64235-6 Pa. $18.95

DE RE METALLICA, Georgius Agricola. The famous Hoover translation of greatest treatise on technological chemistry, engineering, geology, mining of early modern times (1556). All 289 original woodcuts. 638pp. 6¾ x 11.   60006-8 Pa. $21.95

SOME THEORY OF SAMPLING, William Edwards Deming. Analysis of the problems, theory and design of sampling techniques for social scientists, industrial managers and others who find statistics increasingly important in their work. 61 tables. 90 figures. xvii + 602pp. 5⅜ x 8½.   64684-X Pa. $16.95

THE VARIOUS AND INGENIOUS MACHINES OF AGOSTINO RAMELLI: A Classic Sixteenth-Century Illustrated Treatise on Technology, Agostino Ramelli. One of the most widely known and copied works on machinery in the 16th century. 194 detailed plates of water pumps, grain mills, cranes, more. 608pp. 9 x 12.
28180-9 Pa. $24.95

LINEAR PROGRAMMING AND ECONOMIC ANALYSIS, Robert Dorfman, Paul A. Samuelson and Robert M. Solow. First comprehensive treatment of linear programming in standard economic analysis. Game theory, modern welfare economics, Leontief input-output, more. 525pp. 5⅜ x 8½.   65491-5 Pa. $14.95

ELEMENTARY DECISION THEORY, Herman Chernoff and Lincoln E. Moses. Clear introduction to statistics and statistical theory covers data processing, probability and random variables, testing hypotheses, much more. Exercises. 364pp. 5⅜ x 8½.   65218-1 Pa. $10.95

THE COMPLEAT STRATEGYST: Being a Primer on the Theory of Games of Strategy, J.D. Williams. Highly entertaining classic describes, with many illustrated examples, how to select best strategies in conflict situations. Prefaces. Appendices. 268pp. 5⅜ x 8½.   25101-2 Pa. $7.95

CONSTRUCTIONS AND COMBINATORIAL PROBLEMS IN DESIGN OF EXPERIMENTS, Damaraju Raghavarao. In-depth reference work examines orthogonal Latin squares, incomplete block designs, tactical configuration, partial geometry, much more. Abundant explanations, examples. 416pp. 5⅜ x 8¼.
65685-3 Pa. $10.95

THE ABSOLUTE DIFFERENTIAL CALCULUS (CALCULUS OF TENSORS), Tullio Levi-Civita. Great 20th-century mathematician's classic work on material necessary for mathematical grasp of theory of relativity. 452pp. 5⅜ x 8½.
63401-9 Pa. $11.95

VECTOR AND TENSOR ANALYSIS WITH APPLICATIONS, A.I. Borisenko and I.E. Tarapov. Concise introduction. Worked-out problems, solutions, exercises. 257pp. 5⅜ x 8¼.   63833-2 Pa. $8.95

THE FOUR-COLOR PROBLEM: Assaults and Conquest, Thomas L. Saaty and Paul G. Kainen. Engrossing, comprehensive account of the century-old combinatorial topological problem, its history and solution. Bibliographies. Index. 110 figures. 228pp. 5⅜ x 8½.   65092-8 Pa. $7.95

CATALOG OF DOVER BOOKS

CATALYSIS IN CHEMISTRY AND ENZYMOLOGY, William P. Jencks. Exceptionally clear coverage of mechanisms for catalysis, forces in aqueous solution, carbonyl- and acyl-group reactions, practical kinetics, more. 864pp. 5⅜ x 8½.
65460-5 Pa. $19.95

PROBABILITY: An Introduction, Samuel Goldberg. Excellent basic text covers set theory, probability theory for finite sample spaces, binomial theorem, much more. 360 problems. Bibliographies. 322pp. 5⅜ x 8½.
65252-1 Pa. $10.95

LIGHTNING, Martin A. Uman. Revised, updated edition of classic work on the physics of lightning. Phenomena, terminology, measurement, photography, spectroscopy, thunder, more. Reviews recent research. Bibliography. Indices. 320pp. 5⅜ x 8¼.
64575-4 Pa. $8.95

PROBABILITY THEORY: A Concise Course, Y.A. Rozanov. Highly readable, self-contained introduction covers combination of events, dependent events, Bernoulli trials, etc. Translation by Richard Silverman. 148pp. 5⅜ x 8¼.
63544-9 Pa. $7.95

AN INTRODUCTION TO HAMILTONIAN OPTICS, H. A. Buchdahl. Detailed account of the Hamiltonian treatment of aberration theory in geometrical optics. Many classes of optical systems defined in terms of the symmetries they possess. Problems with detailed solutions. 1970 edition. xv + 360pp. 5⅜ x 8½.
67597-1 Pa. $10.95

STATISTICS MANUAL, Edwin L. Crow, et al. Comprehensive, practical collection of classical and modern methods prepared by U.S. Naval Ordnance Test Station. Stress on use. Basics of statistics assumed. 288pp. 5⅜ x 8½.
60599-X Pa. $7.95

DICTIONARY/OUTLINE OF BASIC STATISTICS, John E. Freund and Frank J. Williams. A clear concise dictionary of over 1,000 statistical terms and an outline of statistical formulas covering probability, nonparametric tests, much more. 208pp. 5⅜ x 8½.
66796-0 Pa. $7.95

STATISTICAL METHOD FROM THE VIEWPOINT OF QUALITY CONTROL, Walter A. Shewhart. Important text explains regulation of variables, uses of statistical control to achieve quality control in industry, agriculture, other areas. 192pp. 5⅜ x 8½.
65232-7 Pa. $7.95

METHODS OF THERMODYNAMICS, Howard Reiss. Outstanding text focuses on physical technique of thermodynamics, typical problem areas of understanding, and significance and use of thermodynamic potential. 1965 edition. 238pp. 5⅜ x 8½.
69445-3 Pa. $8.95

STATISTICAL ADJUSTMENT OF DATA, W. Edwards Deming. Introduction to basic concepts of statistics, curve fitting, least squares solution, conditions without parameter, conditions containing parameters. 26 exercises worked out. 271pp. 5⅜ x 8½.
64685-8 Pa. $9.95

TENSOR CALCULUS, J.L. Synge and A. Schild. Widely used introductory text covers spaces and tensors, basic operations in Riemannian space, non-Riemannian spaces, etc. 324pp. 5⅜ x 8¼.
63612-7 Pa. $9.95

CATALOG OF DOVER BOOKS

A CONCISE HISTORY OF MATHEMATICS, Dirk J. Struik. The best brief history of mathematics. Stresses origins and covers every major figure from ancient Near East to 19th century. 41 illustrations. 195pp. 5⅜ x 8½. 60255-9 Pa. $8.95

A SHORT ACCOUNT OF THE HISTORY OF MATHEMATICS, W.W. Rouse Ball. One of clearest, most authoritative surveys from the Egyptians and Phoenicians through 19th-century figures such as Grassman, Galois, Riemann. Fourth edition. 522pp. 5⅜ x 8½. 20630-0 Pa. $11.95

HISTORY OF MATHEMATICS, David E. Smith. Nontechnical survey from ancient Greece and Orient to late 19th century; evolution of arithmetic, geometry, trigonometry, calculating devices, algebra, the calculus. 362 illustrations. 1,355pp. 5⅜ x 8½. 20429-4, 20430-8 Pa., Two-vol. set $26.90

THE GEOMETRY OF RENÉ DESCARTES, René Descartes. The great work founded analytical geometry. Original French text, Descartes' own diagrams, together with definitive Smith-Latham translation. 244pp. 5⅜ x 8½. 60068-8 Pa. $8.95

THE ORIGINS OF THE INFINITESIMAL CALCULUS, Margaret E. Baron. Only fully detailed and documented account of crucial discipline: origins; development by Galileo, Kepler, Cavalieri; contributions of Newton, Leibniz, more. 304pp. 5⅜ x 8½. (Available in U.S. and Canada only) 65371-4 Pa. $9.95

THE HISTORY OF THE CALCULUS AND ITS CONCEPTUAL DEVELOPMENT, Carl B. Boyer. Origins in antiquity, medieval contributions, work of Newton, Leibniz, rigorous formulation. Treatment is verbal. 346pp. 5⅜ x 8½. 60509-4 Pa. $9.95

THE THIRTEEN BOOKS OF EUCLID'S ELEMENTS, translated with introduction and commentary by Sir Thomas L. Heath. Definitive edition. Textual and linguistic notes, mathematical analysis. 2,500 years of critical commentary. Not abridged. 1,414pp. 5⅜ x 8½. 60088-2, 60089-0, 60090-4 Pa., Three-vol. set $32.85

GAMES AND DECISIONS: Introduction and Critical Survey, R. Duncan Luce and Howard Raiffa. Superb nontechnical introduction to game theory, primarily applied to social sciences. Utility theory, zero-sum games, n-person games, decision-making, much more. Bibliography. 509pp. 5⅜ x 8½. 65943-7 Pa. $13.95

THE HISTORICAL ROOTS OF ELEMENTARY MATHEMATICS, Lucas N.H. Bunt, Phillip S. Jones, and Jack D. Bedient. Fundamental underpinnings of modern arithmetic, algebra, geometry and number systems derived from ancient civilizations. 320pp. 5⅜ x 8½. 25563-8 Pa. $8.95

CALCULUS REFRESHER FOR TECHNICAL PEOPLE, A. Albert Klaf. Covers important aspects of integral and differential calculus via 756 questions. 566 problems, most answered. 431pp. 5⅜ x 8½. 20370-0 Pa. $8.95

CATALOG OF DOVER BOOKS

CHALLENGING MATHEMATICAL PROBLEMS WITH ELEMENTARY SOLUTIONS, A.M. Yaglom and I.M. Yaglom. Over 170 challenging problems on probability theory, combinatorial analysis, points and lines, topology, convex polygons, many other topics. Solutions. Total of 445pp. 5⅜ x 8½. Two-vol. set.

Vol. I: 65536-9 Pa. $7.95
Vol. II: 65537-7 Pa. $7.95

FIFTY CHALLENGING PROBLEMS IN PROBABILITY WITH SOLUTIONS, Frederick Mosteller. Remarkable puzzlers, graded in difficulty, illustrate elementary and advanced aspects of probability. Detailed solutions. 88pp. 5⅜ x 8½.

65355-2 Pa. $4.95

EXPERIMENTS IN TOPOLOGY, Stephen Barr. Classic, lively explanation of one of the byways of mathematics. Klein bottles, Moebius strips, projective planes, map coloring, problem of the Koenigsberg bridges, much more, described with clarity and wit. 43 figures. 210pp. 5⅜ x 8½. 25933-1 Pa. $6.95

RELATIVITY IN ILLUSTRATIONS, Jacob T. Schwartz. Clear nontechnical treatment makes relativity more accessible than ever before. Over 60 drawings illustrate concepts more clearly than text alone. Only high school geometry needed. Bibliography. 128pp. 6⅛ x 9¼. 25965-X Pa. $7.95

AN INTRODUCTION TO ORDINARY DIFFERENTIAL EQUATIONS, Earl A. Coddington. A thorough and systematic first course in elementary differential equations for undergraduates in mathematics and science, with many exercises and problems (with answers). Index. 304pp. 5⅜ x 8½. 65942-9 Pa. $8.95

FOURIER SERIES AND ORTHOGONAL FUNCTIONS, Harry F. Davis. An incisive text combining theory and practical example to introduce Fourier series, orthogonal functions and applications of the Fourier method to boundary-value problems. 570 exercises. Answers and notes. 416pp. 5⅜ x 8½. 65973-9 Pa. $11.95

AN INTRODUCTION TO ALGEBRAIC STRUCTURES, Joseph Landin. Superb self-contained text covers "abstract algebra": sets and numbers, theory of groups, theory of rings, much more. Numerous well-chosen examples, exercises. 247pp. 5⅜ x 8½. 65940-2 Pa. $8.95

STARS AND RELATIVITY, Ya. B. Zel'dovich and I. D. Novikov. Vol. 1 of *Relativistic Astrophysics* by famed Russian scientists. General relativity, properties of matter under astrophysical conditions, stars and stellar systems. Deep physical insights, clear presentation. 1971 edition. References. 544pp. 5⅜ x 8½. 69424-0 Pa. $14.95

Prices subject to change without notice.

Available at your book dealer or write for free Mathematics and Science Catalog to Dept. GI, Dover Publications, Inc., 31 East 2nd St., Mineola, N.Y. 11501. Dover publishes more than 250 books each year on science, elementary and advanced mathematics, biology, music, art, literature, history, social sciences and other areas.